Nonlinear Control Techniques for Electro-Hydraulic Actuators in Robotics Engineering

Nonlinear Control Techniques for Electro-Hydraulic Actuators in Robotics Engineering

Qing Guo and Dan Jiang

CRC Press
Taylor & Francis Group
Boca Raton London New York

CRC Press is an imprint of the
Taylor & Francis Group, an **informa** business

CRC Press
Taylor & Francis Group
6000 Broken Sound Parkway NW, Suite 300
Boca Raton, FL 33487-2742

International Standard Book Number-13: 978-1-138-63422-0 (Hardback)

Visit the Taylor & Francis Web site at
http://www.taylorandfrancis.com

and the CRC Press Web site at
http://www.crcpress.com

To my parents

and my daughter.

Qing Guo

Contents

Foreword

I am delighted to introduce the first book on electro-hydraulic system field. When I undertook a project on the Lower Limb Exoskeleton in the year 2010, I was very interested in the electro-hydraulic control system to drive this complicated mechatronics plant. Hence I have tried to investigate many references about the advanced control techniques used in the electro-hydraulic system by many researchers. In these references, especially with the lack of high-cited literature, it indicated that more and more engineers are required to introduce new control ideas into mechanical control plants, not limited to the traditional PI control method. Especially in certain terrible working conditions, perhaps PI control has no favorable dynamic performance or accuracy since its effective model is only two orders. If the control algorithm is not changed, we must select a better sensor or actuator, and other elements to replace the current hardware. This will lead to more cost and consumption of more material. Thus, we need to update the control algorithm and look for better control techniques to improve the performance of the electro-hydraulic system. Based on this concept, now that I have my attention on it, it appears to be a better idea to spread some advanced control techniques applied in the electro-hydraulic control system, especially in typical fields such as robotics engineering and fluid transmission engineering.

Electro-hydraulic servo actuator started gaining recognition in the 1960s as a field. This actuator has many obvious advantages than the general servo motor actuator, such as higher load-bearing and small size-to-power ratio. Many mechatronics plants about multi-DOF manipulator, crane, space manipulator, and lower limb of BigDog favor this actuator due to the existence of large dynamic external load beyond the capability of the servo motor actuator. Furthermore, the electro-hydraulic servo actuator has high control accuracy and fast response performance since the supply pressure can be easily increased. The operation and reliability are also beneficial to the automation control system. In addition, the energy saving of the electro-hydraulic system can also be realized by some pump valve combined control method.

For the researchers of automatic control field, the important purpose is to apply several control methods into practice. However, different actual control plants have different characteristics. There is no generic control method superiorly used in any plant. To address different control problems, the corresponding control scheme should be designed and optimized. Although some control method does not have an ideal performance in special working conditions, engineers cannot give up easily, that is, the issue is analyzed case by case. Thus, in this book, many control methods handle many different control problems and conditions in electro-hydraulic system, which is only for the reference of the reader.

Yan Shi
School of Automation Science and Electrical Engineering
Beihang University
Beijing, China

Preface

Electro-hydraulic servo systems (EHSs) are nowadays widely used in mechatronic control engineering due to their higher load-bearing and small size-to-power ratio. It was found that EHSs are beginning to be commonly applied in some multi-DOF manipulator, crane, space manipulator, and lower limb of BigDog. However, there exist several typical problems that are related to EHS.

One of the fundamental difficulties in electro-hydraulic control is the undesirable dynamic behavior of the designed controller due to the external load disturbance that exists in EHS. This is caused by the driven force or torque of the mechatronics plant. For different types of external load on the electro-hydraulic actuator (EHA), it is often considered as zero [31] or unknown constant [64], even a bounded uncertainty disturbance [7,60,62]. However, Kim et al. [33] developed a high-gain disturbance observer (HGDO) with backstepping control to compensate for the unknown external load while guaranteeing the position tracking error within an acceptable level. Chen [9] proposed a nonlinear disturbance observer integrated with a general nonlinear controller. In practice, the external load may be a largely unknown structural disturbance of EHS, which should be compensated by the constructed controller. Even though the feedback control of EHS may be stable, it is clear that the dynamic performance will be declined if the external load increases beyond the maximum load capability of EHA.

The second problem in electro-hydraulic control system is degraded dynamic and steady-state performance due to the hydraulic parametric uncertainties existing in EHS. To handle this problem, several advanced control methods have also been presented such as local model linearizations [5,38,49], robust H_∞ controllers [20,42], quantitative feedback control schemes [45], geometric control approach [55], output regulation control [52], parametric adaptive controllers [1,18,23,24], state-constrained control [26,27], robust controller with extended state observer [21,61], and disturbance observer [47]. These controllers usually adopted adaptive parametric estimation law (APEL) to estimate the uncertainty parameter. It is to be noted that the load disturbance and parametric uncertainty often lead to unexpected chatter, overshoot, and zero bias of tracking error. Thus, the designed controller should be considered in EHS to guarantee not only the prescribed accuracy of output tracking error but also the desirable dynamic responses of system state.

In addition, as the common nonlinear control system with strict feedback form is usually handled by the well-known backstepping method [37], many high-order derivatives of the virtual control variables are generated in backstepping iteration, which easily result in the derivatives explosion [58], violent control, and saturation. To address this problem, the dynamic surface control (DSC) was proposed to design a stable dynamic surface [54] instead of a virtual control derivative. The advantage of DSC is to eliminate the severe proliferation and singularity of the nonlinear system and guarantee fast state convergence and desirable dynamic performance [50]. The dynamic surface is often designed as a linear filter to transform high-order derivations of virtual control variables into a different stable dynamic surface. If the DSC is not considered in the backstepping controller, Guo et al. [18] proposed another computation method of virtual control variable to avoid the derivatives explosion. This virtual control variable can be directly filtered by a linear decayed memory filter to smooth the high-order derivatives.

In this book, based on the aforementioned contributions of many researchers, the authors have tried their best to apply some typical linear and nonlinear control techniques into EHAs, which drive a two-DOF robotic arm, for example. Some control ideas and motion mechanism are partially similar to many references. Both theoretical proof and simulation and experimental results are given to verify the corresponding control method in detail. The authors hope that these points of view can benefit the reader to study the electro-hydraulic control system in depth.

List of Figures

List of Tables

Acknowledgement

Qing Guo and Dan Jiang acknowledge Professor Andrew Plummer, director of Power Center for Power Transmission and Motion Control, Department of Mechanical Engineering, University of Bath for the funding assistance and previous work. They thank Dr. Tian Yu and Dr. Can Du from Power Center for Power Transmission and Motion Control for their collaborative study in related electrohydraulic work.

During the writing of the book, the authors were assisted by many researchers from electrohydraulic engineering regarding some innovative methods and techniques. Although some theoretical methods have a long way to go in the application of electrohydraulic engineering, scholars did not give up using new control methods and techniques, which is much appreciated by them.

The study was supported by UK Engineering and Physical Sciences Research Council project (No. EP/H024190/1), the National Natural Science Foundation of China (Nos. 61305092 and 51205045), the Postdoctoral Science Foundation of China (No. 2013M542487), the Fundamental Research Funds for the Central Universities, China (Nos. 20160250331 and ZYGX2016J160), and the Open Foundation of the State Key Laboratory of Fluid Power & Mechatronic Systems (No. GZKF-201515).

Authors

Qing Guo earned his B.E. in automation from Harbin Institute of Technology (Harbin, China) in 2003, and went on to earn his M.S. and Ph.D. there, in 2005 and 2008, respectively. In 2009, he became a lecturer at the School of Aeronautics and Astronautics, University of Electronic Science and Technology of China (ChengDu, China), and in 2013, he was promoted to associate professor. From December 2013 to December 2014, he served as an academic visitor at the Center for Power Transmission and Motion Control, Department of Mechanical Engineering, University of Bath, United Kingdom. Dr. Guo's research interests include robust and adaptive control, mechatronic systems, and rehabilitation robots.

Dan Jiang earned her B.E. in mechanical engineering (2002), and M.S. (2005) and Ph.D. (2009) in fluid power transmission and control from the Harbin Institute of Technology (Harbin, China). Since April 2009, she has been with the School of Mechatronics Engineering, University of Electronic Science and Technology of China (ChengDu, China) as a lecturer. Her research interests include hydraulic control and microfluidic technology, and she has published several research papers in these areas.

Symbols

Two-DOF	Two-degree-of-freedom
EHS	Electro-hydraulic system
EHA	Electro-hydraulic actuator
K_{sv}	Gain of the servo valve
T_{sv}	Time constant of the servo valve
u	Input control voltage of the servo valve
C_d	Discharge coefficient
w	Area gradient of the servo valve spool
p_L	Load pressure of the servo valve
p_s	Supply pressure
x_v	Spool position of the servo valve
ρ	Density of the hydraulic oil
y	Displacement of the cylinder
y_d	Demand displacement
C_{tl}	Total leakage coefficient of the cylinder
β_e	Effective bulk modulus
A_p	Annulus area of the symmetrical chamber
A_a	Annulus area of the single-rod chamber
A_b	Annulus area of the no-rod chamber
V_t	Half-volume of the cylinder
m	Load mass
K	Load spring constant
b	Viscous damping coefficient
F_L	External load of the hydraulic actuator
k	Positive constant in the hyperbolic tangent function
sgn(.)	The sign function
tanh(.)	The hyperbolic tangent function
z_i	System state error
x_i	System state
ϑ_i	Parametric uncertainty
$\hat{\vartheta}_i$	Parametric estimation
$\tilde{\vartheta}_i$	Parametric estimation error
θ_i	Joint angle of the robotic arm
d	Equivalent disturbance
$\hat{d}$	Disturbance estimation
$\tilde{d}$	Disturbance estimation error

1

Introduction

CONTENTS

In mechatronic engineering, there exist three common actuators used in automation system, that is, servo motor, electro-hydraulic, and pneumatic. The servo motor system has high mechanical efficiency and control accuracy than the other two actuators. However, its power is restricted by the element size and structure of motor and driver. Pneumatic actuator is an environmentally friendly actuator, which has no obvious pollution to environment. But the pneumatic actuator has very low execution efficiency in engineering. Many available energy dissipates as heat. Although the mechanical efficiency of the electro-hydraulic actuator (EHA) is not higher than that of the servo motor actuator, electro-hydraulic servo systems (EHSs) are nowadays more widely used in mechatronic control engineering due to their higher load-bearing and small size-to-power ratio [40] than the general servo motor actuator. When the supply pressure is large, the dynamic response may also be faster than the servo motor under some large external load and other unknown disturbance. It was found that EHSs are beginning to be commonly applied for large power equipment such as wheel loaders [13], load simulators [63], insulator fatigue test devices [66], and exoskeletons [22]. EHA is a favorable control executive part to realize some multi-DOF manipulator, crane, space manipulator, and lower limb of BigDog due to the large dynamic external load.

In this book, four control methods are discussed in the electro-hydraulic system such as linear classical control, robust control, nonlinear control, and adaptive control methods. Each method has its own advantages to address different control problems. The authors have investigated numerous references about the hydraulic control of EHS. Generally, four typical problems need to be addressed in EHS control (Figure 1.1). The first problem is parametric uncertainties such as effective bulk modulus β_e, leakage coefficient of the cylinder C_{tl}, and hydraulic oil density ρ. These parameters are often unknown constants in different working conditions. To solve the unknown constant parameters, the parametric adaptive estimation (PAE) law is adopted to estimate some parametric uncertainties. The second problem is unmodeled uncertainties, such as nonlinear friction of cylinder and viscous resistance of hydraulic oil [3,17]. These parameters are dynamic variables, which are often adopted by extended state observer (ESO) [61] or disturbance observer (DO) [59]. The third problem is output feedback control, which assumes that some system states are not directly measured from sensors. Thus, we need to construct the high-gain observer [33–35] to estimate some unmeasured system states, which are used in the

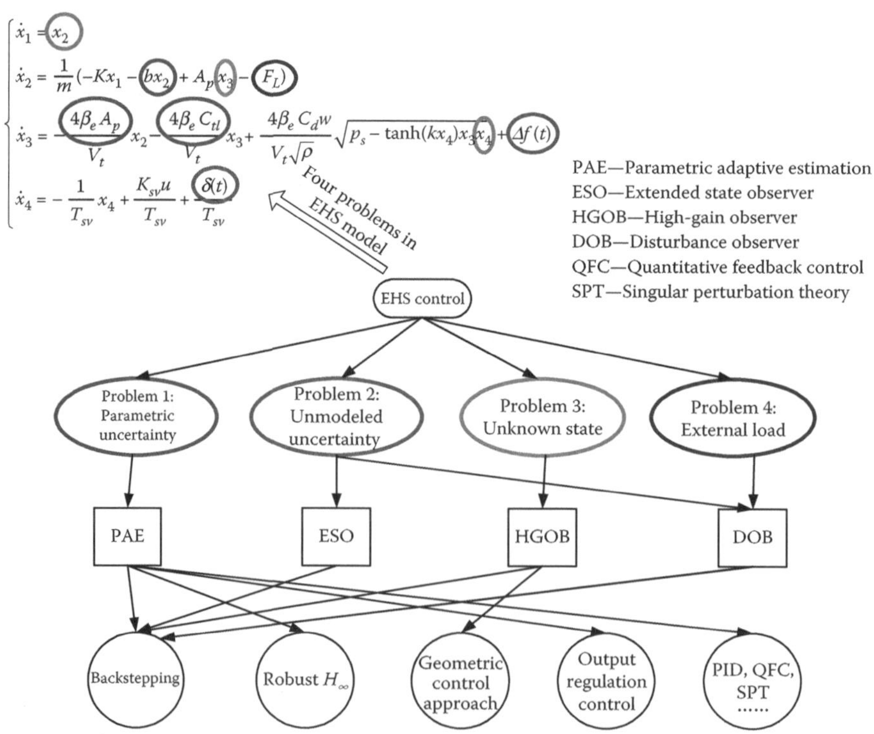

FIGURE 1.1
Some typical problems and the corresponding solutions in the EHS model.

controller design. The last problem is the external load on the EHA, which is a largely unknown dynamic variable. To address the external load, a simple method is adopted by the DO. In this book, we mainly consider Problems 1 and 4, that is, parametric uncertainty and external load disturbance. These two problems may degrade the dynamic behavior and steady-state control accuracy of EHS and the robustness of the designed controller.

1.1 Parametric Uncertainty Problem of EHS

Since some unknown parametric variation may be significant in different working conditions (i.e., phenomenon such as oil temperature variations, pressure-flow characteristics, hysteresis in flow gain characteristics, oil leakage, characteristics of valves near null) [42], many state/output feedback controllers cannot be well established to guarantee the dynamic performance of EHS. Thus, the parametric estimation is one available method to obtain unknown parameters by state observer construction integrated with other nonlinear controller. However, different from parametric uncertainties, the main disturbance is the largely unknown torque/-force disturbance caused by external loadings on the hydraulic actuator. So, further references have been more focused on disturbance rejection of EHS. For instance, Reference 64 proposed a nonlinear controller in which the external load is treated as an uncertain but bounded disturbance. It has been shown that the closed-loop stability can be directly analyzed by Lyapunov technique. Yao and Bu [7,60] assumed that the maximum relative uncertainty of the external load disturbance is bounded by a known value and proposed a discontinuous projection-based adaptive backstepping controller. Kim et al. [33] presented a DO with proportional integral (PI) control form to estimate a biased sinusoidal external load. Then, Won et al. [59] developed a high-gain disturbance observer (HGDO) with backstepping to compensate for the unknown external load and guaranteed tolerance of the position tracking error. These references denote that the external load with wide variations is an important factor to decline the dynamic response performance of the hydraulic controller, especially in some critical condition where the external load of the hydraulic actuator is close to the limitation. Therefore, to improve the dynamic behavior of EHS, various advanced control approaches were developed to estimate unknown parametric uncertainties and unmeasured disturbance.

Recently, some robust H_∞ control methods [19,20,28,29] and quantitative feedback theories [45] have been presented to overcome parametric uncertainties and to guarantee the robustness of the controller. A geometric control approach [55] was verified by numerical simulation to realize the dynamic tracking position of a single-rod cylinder. In the past two decades, backstepping control was widely used in EHS [2,3,56]. If the model of EHS is a strict feedback system [32,37], the backstepping controller based on state feedback can be well implemented. Bu and Yao [8] presented a discontinuous projection-based adaptive backstepping controller to estimate some unknown parameters of the asymmetric hydraulic actuator. Subsequently, Guan and Pan [16,17] also constructed a parametric adaptive estimation law to guarantee the asymptotic convergence of the backstepping controller. Kaddissi et al. [31] proposed an equivalent parameter identification method by the least squares and obtained better performance than the linear controller. Ahn et al. [1] presented an adaptive position control for a pump-controlled EHA based on an adaptive backstepping control scheme. To address some unmeasured physical states, Sun and Chiu [53] proposed a perturbation observer to estimate the load pressure of a single-rod hydraulic actuator. Pi and Wang [48] designed an observer-based cascade controller to estimate the disturbance force in the hydraulic manipulator. In addition, some output observers [34,35,44,61] were used to estimate hydraulic states with less measured information than state feedback observers. These various observers were verified in backstepping or other nonlinear control method.

1.2 Largely Unknown Load Disturbance of EHS

The other difficulty in EHS is the undesirable dynamic behavior of the established controller due to the ignorance of the largely unknown disturbance caused by the dynamic external load (i.e., torque/force). As far as the authors know, the external load was not well addressed in aforementioned studies. If the external load disturbance is treated as zero [56] or a known constant [1], even bounded by a known value [64], many novel controllers are designed conveniently by Lyapunov technique. Thus, Yao and Bu [7,60] assumed that the maximum relative uncertainty of the external load disturbance was bounded by a known value and proposed a discontinuous projection-based adaptive backstepping controller. Yao et al. [62] presented an adaptive robust controller to handle the nonlinear parametric uncertainty in an auxiliary function. In this approach, the first and second derivatives of the external load disturbance were bounded by two constants directly in order to conveniently obtain the negative definite Lyapunov function. Then, he also assumed that the modeling uncertainty and the external load disturbance became zero after a finite time in Reference 61. Kaddissi et al. [31] proposed an equivalent parameter identification method by the least squares and obtained better performance than the linear controller. In this method, the load torque disturbance was still assumed as zero. However, in engineering practice, the external load is often the largely unknown disturbance of EHS. Thus, the designed controller would not only eliminate system state error but also suppress the unknown disturbance of the external load. Even though the EHS may be stable, it is clear that the dynamic performance will be declined if the external load increases beyond a definite boundary [25].

Subsequently, Kim et al. [36] proposed a flatness-based nonlinear controller to improve the position tracking performance while assuming the known constant external load. Then his research team developed a HGDO with backstepping control to compensate for the unknown external load and guaranteed the position tracking accuracy. In this novel approach, the DO had two different forms. One was a second-order high-pass filter [33] to estimate a sinusoidal disturbance with unknown frequency. The other was a HGDO [59] to estimate the largely unknown disturbance caused by the friction, the load force, and the parameter uncertainties. Both the simulation and the experimental results indicated that the external load can be well compensated, as well as the extended system state error was ultimate boundedness. Chen proposed a nonlinear DO integrated with general nonlinear controller [9]. In practice, the external load may be the largely unknown structural disturbance of EHS, which should be compensated by the constructed controller.

1.3 Control Method Illustration in EHS

Recently, there have been several control methods applied in electro-hydraulic control system. The classical approach to the control of EHSs is proportional integral derivative (PID) control, which is easy to be used in industry. Variable control parameters of PID controllers [6,43] are adopted to suit the variable characteristics of the dynamic model. The advantage of the linear PID controller is the simple structure and the feedback variable is only the output information, not all the state or other indirect computed variables. However, the linear controller cannot effectively compensate the external load and has an obvious bias error that exists in tracking response due to the parametric uncertainty. Thus, some advanced controllers are used in EHS to improve the dynamic performance under the uncertain hydraulic parameters and load disturbances. The robust controller can improve the robustness of the feedback control system according to small gain theorem. But the controller is relatively conservative. The parametric adaptive control method often addresses the parametric uncertainty and has an obvious effect of the parametric estimation. However, some unmodeled uncertainties are not well addressed due to the unknown boundness of the hydraulic parameter. Furthermore, the nonlinear model of EHS is assumed to be strict feedback form. Otherwise, the backstepping control method cannot be used to derive the virtual control variable.

The feedback control method is often of two forms, that is, the state feedback and the output feedback. The former needs adequate sensors to measure the state of EHS. In practice engineering, due to the cost constraints and measurement reliability of engineering, many hydraulic states may not be easy to obtain, such as pressures in different cylinder chambers, spool position, and its velocity. Therefore, the output feedback control method should be investigated to achieve a similar performance to the state feedback control. Some output state observers are used in EHS to estimate hydraulic states, which can be used in backstepping or other nonlinear control method [34,35,44]. These state observers are designed as proportional or PI structure. If the observer is convergence, the dynamic behavior can be obviously improved. Apart from the above-mentioned observer, Sun presented a perturbation observer to estimate the different chamber pressure of the single-rod hydraulic actuator [53]. Pi and Wang constructed a DO to estimate and compensate the unknown disturbance in the hydraulic manipulator [48].

2

Model Construction of Electro-Hydraulic Control System

CONTENTS

The general electro-hydraulic system model often includes four elements such as hydraulic cylinder, servo valve, mechatronics plant, and hydraulic parametric uncertainty. In terms of the model construction of the electro-hydraulic actuator, the mechatronics plant can be omitted and is usually replaced by the so-called external load disturbance. In this chapter, these model elements are illustrated in detail.

2.1 Hydraulic Cylinder Model

In EHS, the hydraulic cylinder is one type of motion actuator to realize the linear stretch of the plant. It often has two different styles, that is, symmetrical hydraulic cylinder and asymmetrical hydraulic cylinder. The former is a double-rod acting mechanism in both sides of the cylinder and the latter is a single-rod acting mechanism in only one side of the cylinder as shown in Figure 2.1.

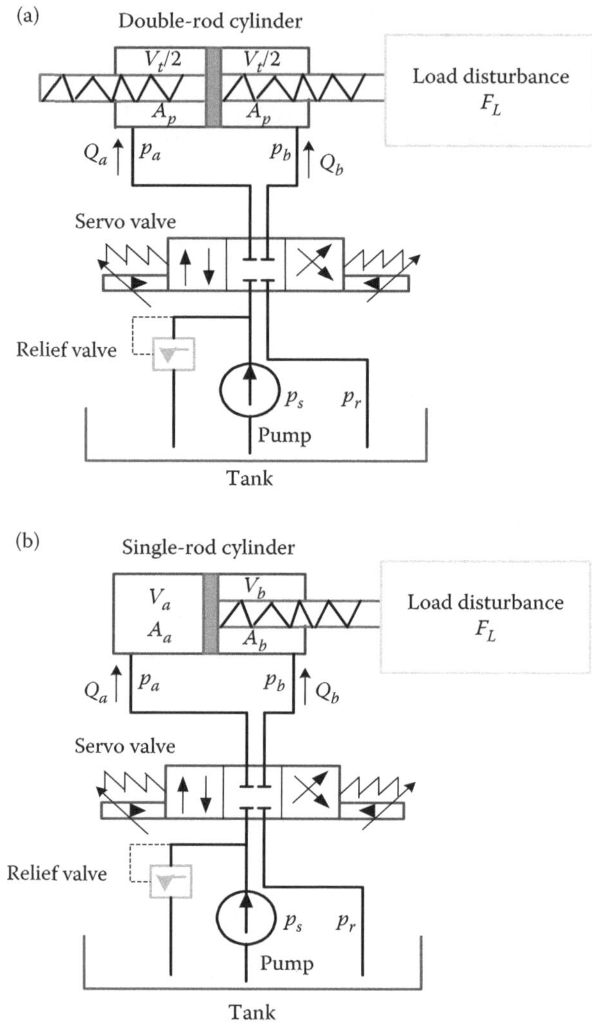

FIGURE 2.1
Symmetrical and asymmetrical hydraulic cylinders: (a) the double-rod acting mechanism of the symmetrical hydraulic cylinder and (b) the single-rod acting mechanism of the asymmetrical hydraulic cylinder.

2.1.1 Symmetrical Cylinder Model

By referring to the symmetrical cylinder model mentioned in References 40 and 41, the flow-pressure continuous equation of the hydraulic cylinder is

$$Q_L = A_p\dot{y} + C_{tl}p_L + \frac{V_t}{4\beta_e}\dot{p}_L, \tag{2.1}$$

where $p_L = p_a - p_b$ is the load pressure, Q_L is the load flow, y is the displacement of the cylinder, C_{tl} is the coefficient of the total leakage of the cylinder, β_e is the effective bulk modulus, A_p is the annulus area of the cylinder chamber, and V_t is the half-volume of the cylinder.

When the cylinder starts to move, the spring force and the viscous friction of hydraulic oil are two typical resistances of the electro-hydraulic system. Especially in load condition, the hydraulic pressure has to overcome the external load force to drive the plant motion. Thus, if the major viscous friction of hydraulic oil is simplified as coulomb friction, the mechanical dynamic equation driven by the hydraulic actuator can be constructed as follows:

$$m\ddot{y} = p_L A_p - Ky - b\dot{y} - F_L, \tag{2.2}$$

where m is the load mass, K is the load spring constant, b is the viscous damping coefficient, and F_L is the external load on the hydraulic actuator.

2.1.2 Asymmetrical Cylinder Model

The flow-pressure continuous model of the hydraulic cylinder is given by

$$\begin{cases} A_a\dot{y} + C_{tl}(p_a - p_b) + (V_{0a} + A_a y)\dot{p}_a/\beta_e = Q_a \\ A_b\dot{y} + C_{tl}(p_a - p_b) - (V_{0b} - A_b y)\dot{p}_b/\beta_e = Q_b \end{cases}, \tag{2.3}$$

where Q_a and Q_b are load flows with the spool position of the servo valve $x_v \geq 0$ and $x_v < 0$, respectively, p_a and p_b are the pressures inside the two chambers of the cylinder, A_a and A_b are the ram areas of the two chambers, and $V_{0a} = V_t/2$ and $V_{0b} = V_t/2$ are the initial total control volumes of the two cylinder chambers, respectively.

Different from the symmetrical cylinder, the load pressure of the asymmetrical cylinder $p_L = (p_a A_a - p_b A_b)/A_a$, the mechanical dynamic equation is described as follows:

$$m\ddot{y} = p_a A_a - p_b A_b - Ky - b\dot{y} - F_L(t). \tag{2.4}$$

2.2 Servo Valve Model

The servo valve model includes the load flow model and the spool position response model. In terms of the zero-opened four-way spool valve, the load flow of the valve can be described differently for symmetrical and asymmetrical cylinders, respectively.

2.2.1 Load Flow Model

For the symmetrical cylinder, the load flow of the valve is simplified as follow:

$$Q_L = C_d w x_v \sqrt{\frac{1}{\rho}(p_s - \text{sgn}(x_v) p_L)}, \tag{2.5}$$

where x_v is the spool position of the servo valve, p_s is the supply pressure of the pump, C_d is the discharge coefficient, w is the area gradient of the servo valve spool, and ρ is the density of the hydraulic oil, and sgn(.) is the sign function, that is, $\text{sgn}(x) = 1$, if $x > 0$, else if $x < 0$, $\text{sgn}(x) = -1$, else $\text{sgn}(x) = 0$.

For the asymmetrical cylinder, the load flow of the valve cannot be described as Q_L, and the flow in the two directions Q_a and Q_b is given by

$$\begin{aligned} Q_a &= \begin{cases} C_d w x_v \sqrt{2(p_s - p_a)/\rho} & x_v \geq 0 \\ C_d w x_v \sqrt{2(p_a - p_r)/\rho} & x_v < 0 \end{cases}, \\ Q_b &= \begin{cases} C_d w x_v \sqrt{2(p_b - p_r)/\rho} & x_v \geq 0 \\ C_d w x_v \sqrt{2(p_s - p_b)/\rho} & x_v < 0 \end{cases}, \end{aligned} \tag{2.6}$$

where Q_a is the main load flow as $x_v \geq 0$, Q_b is the main load flow $x_v < 0$, and p_r is the return pressure of the tank.

2.2.2 Spool Position Response Model

The spool position response model describes the relationship between the spool position and the input control voltage. Without loss of generality, this model can be constructed as three forms, that is, a simplified model, first-order linear dynamic model, and second-order dynamic model.

For the simplified model, since the cut-off frequency of the servo valve is far greater than the control system bandwidth, the valve dynamics can be neglected in model construction [33]. Thus, the spool position response model is given by

$$x_v = K_{sv} u, \tag{2.7}$$

where K_{sv} is the gain of the servo valve and u is the control voltage of the servo valve.

For the one-order linear dynamic model, it is considered one time constant in the spool position response to denote the command delay of the servo valve, which is constructed as follows:

$$T_{sv} \dot{x}_v + x_v = K_{sv} u, \tag{2.8}$$

where T_{sv} is the response time constant of the servo valve.

If both the delay and damping characteristics are considered in the spool position response model of the servo valve, then the second-order dynamic model is given by

$$\ddot{x}_v + 2\zeta_{sv}\omega_{sv}\dot{x}_v + \omega_{sv}^2 x_v = K_{sv}\omega_{sv}^2 u, \tag{2.9}$$

where ζ_{sv} is the damping ratio and ω_{sv} is the natural frequency.

2.3 Parametric Uncertainty and Load Disturbance

Electro-hydraulic systems often have some model uncertainties such as parametric uncertainty and external load disturbance. The former is mostly due to unknown viscous damping, load stiffness, variations in control fluid volumes, physical characteristics of the valve, bulk modulus, and oil temperature variations [13], and the latter is caused by the driven force or torque of the mechatronic plant. However, in practice, all the parametric uncertainty and load disturbance are unknown. Without loss of generality, some hydraulic parametric uncertainties are considered to be bounded by several unknown or known constants as follows:

$$\begin{aligned} &C_d = \bar{C}_d + \Delta C_d, \quad w = \bar{w} + \Delta w, \quad \beta_e = \bar{\beta}_e + \Delta\beta_e, \\ &C_{tl} = \bar{C}_{tl} + \Delta C_{tl}, \quad \rho = \bar{\rho} + \Delta\rho, \quad V_t = \bar{V}_t + \Delta V_t, \end{aligned} \tag{2.10}$$

where $\bar{C}_d$, $\bar{w}$, $\bar{\beta}_e$, $\bar{C}_{tl}$, $\bar{\rho}$, and $\bar{V}_t$ are known normal hydraulic parameters, and their uncertainties are constrained as follows:

$$\begin{aligned} &\Delta C_{d\min} \le \Delta C_d \le \Delta C_{d\max}, \quad \Delta w_{\min} \le \Delta w \le \Delta w_{\max}, \\ &\Delta\beta_{e\min} \le \Delta\beta_e \le \Delta\beta_{e\max}, \quad \Delta C_{tl\min} \le \Delta C_{tl} \le \Delta C_{tl\max}, \\ &\Delta\rho_{\min} \le \Delta\rho \le \Delta\rho_{\max}, \quad \Delta V_{t\min} \le \Delta V_t \le \Delta V_{\max}. \end{aligned} \tag{2.11}$$

As shown in Equation 2.11, $\Delta C_{d\min}$, $\Delta C_{d\max}$, $\Delta w_{\min}$, $\Delta w_{\max}$, $\Delta\beta_{e\min}$, $\Delta\beta_{e\max}$, $\Delta C_{tl\min}$, $\Delta C_{tl\max}$, $\Delta\rho_{\min}$, $\Delta\rho_{\max}$, $\Delta V_{t\min}$, and $\Delta V_{t\max}$ are unknown or known boundaries of parametric uncertainties.

Remark 2.1

The above parametric uncertainties should be estimated in controller design. Otherwise, the dynamic performance of the electro-hydraulic system will be declined and the robustness of the designed controller will also be degraded.

The external load F_L is considered to be a structural disturbance of the electro-hydraulic system. Since F_L is caused by the driven force or torque of the mechatronic plant, it is reasonable to assume that the dynamic value of F_L

depends on the motion position, velocity, and acceleration of the hydraulic cylinder, that is, $y, \dot{y}, \ddot{y}$. Without loss of generality, F_L can be described as the following two forms.

The first form of $F_L(t)$ is bounded by

$$|F_L(t)| \leq c_0 \left|y\right| + c_1 \left|\dot{y}\right| + c_2 \left|\ddot{y}\right|, \tag{2.12}$$

where c_0, c_1, c_2 are known or unknown weighting boundaries.

Remark 2.2

By Equation 2.12, the boundary discussion of $F_L(t)$ can be transformed into the direct state dynamics of y. Thus, the stability analysis of the electro-hydraulic system model and the control design become easy.

On the other hand, the external load F_L is constrained by

$$|F_L(t)| \leq \Delta_{F_L}, \tag{2.13}$$

where Δ_{F_L} is the known and unknown boundary constant.

Remark 2.3

The advantage of Equation 2.13 is very clear to denote the constraint of load disturbance. However, if the boundary Δ_{F_L} is unknown, the DO should be used to estimate Δ_{F_L} rather than to estimate the dynamic unknown variable $F_L(t)$. Even though Δ_{F_L} is a known value, the controller is relatively conservative as Δ_{F_L} is discussed as a direct uncertainty constraint in control design.

2.4 Nonlinear State-Space Model

According to Equations 2.1, 2.2, and 2.7, the simplest electro-hydraulic actuator model without the spool position response model is given by

$$\begin{cases} \dot{x}_1 = x_2 \\ \dot{x}_2 = \dfrac{1}{m}(-Kx_1 - bx_2 + A_p x_3 - F_L(t)) \\ \dot{x}_3 = -\dfrac{4\beta_e A_p}{V_t} x_2 - \dfrac{4\beta_e C_{tl}}{V_t} x_3 + \dfrac{4\beta_e C_d w K_{sv}}{V_t \sqrt{\rho}} \sqrt{p_s - \mathrm{sgn}(u) x_3} u \end{cases}, \tag{2.14}$$

where $[x_1, x_2, x_3]^T = [y, \dot{y}, p_L]^T$ are three state variables of the electro-hydraulic actuator and u is the control variable of the servo valve.

If the one-order spool position response model is considered, then the electro-hydraulic actuator model is constructed as follows:

$$\begin{cases} \dot{x}_1 = x_2 \\ \dot{x}_2 = \dfrac{1}{m}(-Kx_1 - bx_2 + A_p x_3 - F_L(t)) \\ \dot{x}_3 = -\dfrac{4\beta_e A_p}{V_t}x_2 - \dfrac{4\beta_e C_{tl}}{V_t}x_3 + \dfrac{4\beta_e C_d w}{V_t\sqrt{\rho}}\sqrt{p_s - \text{sgn}(x_4)x_3}x_4 \\ \dot{x}_4 = -\dfrac{1}{T_{sv}}x_4 + \dfrac{K_{sv}}{T_{sv}}u \end{cases}, \tag{2.15}$$

where $[x_1, x_2, x_3, x_4]^T = [y, \dot{y}, p_L, x_v]^T$ are four state variables of the electro-hydraulic actuator.

If the asymmetrical cylinder model is considered with the second-order spool position response model, then, according to Equations 2.3, 2.4, 2.6, and 2.9, the electro-hydraulic actuator model is given by

$$\begin{cases} \dot{x}_1 = x_2 \\ \dot{x}_2 = (x_3 A_a - x_4 A_b - Kx_1 - bx_2 - F_L)/m \\ \dot{x}_3 = h_1\left(-A_a x_2 - C_{tl}(x_3 - x_4)\right) \\ \qquad + h_1 C_d w x_5 \sqrt{2/\rho}\left(s_1\sqrt{p_s - x_3} + s_2\sqrt{x_3 - p_r}\right) \\ \dot{x}_4 = h_2\left(A_b x_2 + C_{tl}(x_3 - x_4)\right) \\ \qquad - h_2 C_d w x_5 \sqrt{2/\rho}\left(s_1\sqrt{x_4 - p_r} - s_2\sqrt{p_s - x_4}\right) \\ \dot{x}_5 = x_6 \\ \dot{x}_6 = -2\zeta_{sv}\omega_{sv}x_6 - \omega_{sv}^2 x_5 + K_{sv}\omega_{sv}^2 u \end{cases}, \tag{2.16}$$

where $[x_1, x_2, x_3, x_4, x_5, x_6]^T = [y, \dot{y}, p_a, p_b, x_v, \dot{x}_v]^T$, and

$$\begin{aligned} h_1 &= \frac{\beta_e}{V_{0a} + A_a x_1}, \quad h_2 = \frac{\beta_e}{V_{0b} - A_b x_1}, \\ s_1 &= \frac{1 + \text{sgn}(x_5)}{2}, \quad s_2 = \frac{1 - \text{sgn}(x_5)}{2}. \end{aligned} \tag{2.17}$$

Remark 2.4

The above three models (2.14), (2.15), and (2.16) are all used in practical control design. The difference is the model accuracy. In conventional condition, the model (2.14) is enough to describe the dynamics of the electro-hydraulic system. However, in largely unknown load disturbance and parametric uncertainty, the models (2.15) and (2.16) are necessary to be used in advanced control method.

2.5 Linearized Model of Electro-Hydraulic Actuator

The linearized model of the electro-hydraulic actuator is necessary to be used in the linear controller design such as PID, which is very widely adopted in engineering (Figure 2.2). For the symmetrical cylinder, the linearized model is derived from Equations 2.1, 2.2, 2.5, and 2.8.

At first, the load flow Q_L in Equation 2.5 is linearized as follows:

$$Q_L = K_q x_v + K_c p_L, \tag{2.18}$$

where K_q and K_c are linearized coefficients from the derivatives of Equation 2.5 as follows:

$$\begin{aligned} K_q &= \frac{\partial Q_L}{\partial x_v} = C_d w \sqrt{\frac{p_s - p_L}{\rho}}, \\ K_c &= \frac{\partial Q_L}{\partial p_L} = \frac{C_d w x_v}{2\sqrt{\rho(p_s - p_L)}}. \end{aligned} \tag{2.19}$$

Second, from Equations 2.1, 2.2, 2.19, and 2.8, we can describe the linearized model of the electro-hydraulic actuator as follows:

Thus, from Figure 2.2, the linearized model of the double-rod electro-hydraulic actuator is given by

$$y(s) = \frac{\frac{K_q K_{sv}}{A_p(T_{sv}s+1)} u(s) - \frac{1}{A_p^2}\left(K_{ce} + \frac{V_t}{4\beta_e}s\right) F_L(s)}{\frac{V_t m}{4\beta_e A_p^2}s^3 + \left(\frac{m_t K_{ce}}{A_p^2} + \frac{B_e V_t}{4\beta_e A_p^2}\right)s^2 + \left(1 + \frac{bK_{ce}}{A_p^2} + \frac{V_t K}{4\beta_e A_p^2}\right)s + \frac{KK_{ce}}{A_p^2}}, \tag{2.20}$$

where $y(s)$, $u(s)$, and $F_L(s)$ are frequency domain transformations, and $K_{ce} = K_c + C_{tl}$.

In terms of the single-rod electro-hydraulic actuator, the load pressure p_L is defined as $p_L = p_a - \upsilon p_b$, and the load flow Q_L is defined as $Q_L = (Q_a + \upsilon Q_b)/(1 + \upsilon^2)$, where υ is the annulus area ratio between the

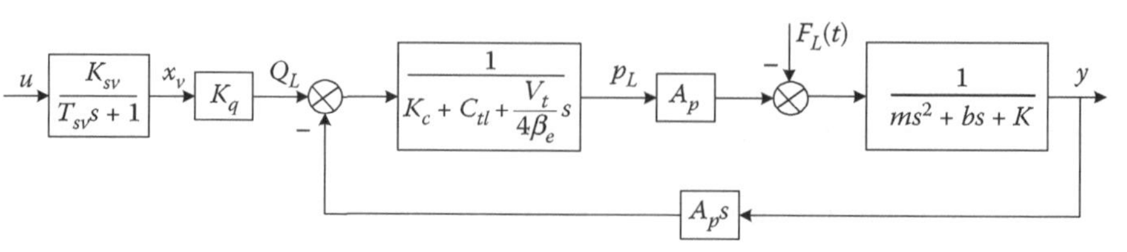

FIGURE 2.2
Linearized model of the electro-hydraulic actuator.

cylinder with rod and the cylinder without rod. Thus, the load flow Q_L is linearized as follows:

$$Q_L = K_{qi}x_v - K_{ci}p_L, \quad i = a, b. \tag{2.21}$$

From Equation 2.6, we can obtain that

$$Q_L = \begin{cases} C_d w x_v \sqrt{\dfrac{2(p_s - p_L)}{\rho(1+\upsilon^3)}}, & x_v \geq 0 \\ C_d w x_v \sqrt{\dfrac{2(np_s + p_L)}{\rho(1+\upsilon^3)}}, & x_v < 0 \end{cases}. \tag{2.22}$$

The linearized coefficients in Equation 2.21 can be derived from Equation 2.22 as follows:

$$\begin{aligned} K_{qa} &= C_d w \sqrt{\frac{2(p_s - p_L)}{\rho(1+\upsilon^3)}}, \quad K_{ca} = \frac{C_d w x_v}{\sqrt{2\rho(p_s - p_L)(1+\upsilon^3)}}, \\ K_{qb} &= C_d w \sqrt{\frac{2(\alpha p_s + p_L)}{\rho(1+\upsilon^3)}}, \quad K_{cb} = -\frac{C_d w x_v}{\sqrt{2\rho(\upsilon p_s + p_L)(1+\upsilon^3)}}. \end{aligned} \tag{2.23}$$

Then, like Equation 2.20, the linearized model of the single-rod electro-hydraulic actuator is also obtained as follows:

$$y(s) = \frac{K_{qi}K_{sv}u(s) - \frac{T_{sv}s+1}{A_a}\left(K_{cei} + \frac{V_t}{2(1+\upsilon^2)\beta_e}s\right)F_L(s)}{Den}, \tag{2.24}$$

where $i = a, b$, $K_{cei} = K_{ci} + C_{tl}$, and

$$\begin{aligned} Den = & A_a(T_{sv}s+1)\left(\frac{V_t m}{2(1+\upsilon^2)\beta_e A_a^2}s^3 + \left(\frac{mK_{cei}}{A_a^2} + \frac{bV_t}{2(1+\upsilon^2)\beta_e A_a^2}\right)s^2\right. \\ & \left. + \left(1 + \frac{bK_{cei}}{A_a^2} + \frac{V_t K}{2(1+\upsilon^2)\beta_e A_a^2}\right)s + \frac{KK_{cei}}{A_a^2}\right). \end{aligned} \tag{2.25}$$

Remark 2.5

The two linearized models (2.20) and (2.24) can be used in classical control design such as PID, robust H_∞, where a linear controller is obtained. This linear controller is relatively simple to guarantee the stability of the closed-loop system and basic dynamic behavior. However, there exist unknown parametric uncertainty and load disturbance in the electro-hydraulic system, which lead to the existing model error in linearized models. Thus, the two nonlinear models (2.15) and (2.16) are adopted in the nonlinear controller design. Many nonlinear controllers such as backstepping, adaptive, slide mode, etc. should guarantee that the transient and steady behavior of the electro-hydraulic system achieve the prescribe performance. Furthermore, the designed controller must guarantee the global convergence or ultimate boundedness of the generalized system state, including hydraulic state variable, parametric estimation error, and DO error.

3

Linear PID Control Design

CONTENTS

The linear PID control is a classical control method, which is very easy to be understood. The PID controller is designed in frequency domain based on the linear model of the electro-hydraulic system. The main evaluation indexes for the linear PID control design are relative stability margin and bandwidth. This design idea is derived from the frequency design method in classical control principle. In this chapter, the linear PID control design method is introduced and the system stability and dynamic behavior are discussed.

3.1 Linear Feedback Control Loop

According to Figure 2.2, the whole linear feedback control loop of the electro-hydraulic system is shown in Figure 3.1. This is a typical cylinder position feedback control loop. The actual cylinder position Y is controlled by the PID controller to track the position demand Y_{exp}. The servo valve and the cylinder model can be considered as the plant model, which is similar to the linearized model of the double-rod electro-hydraulic actuator (2.20). Here, the dynamics

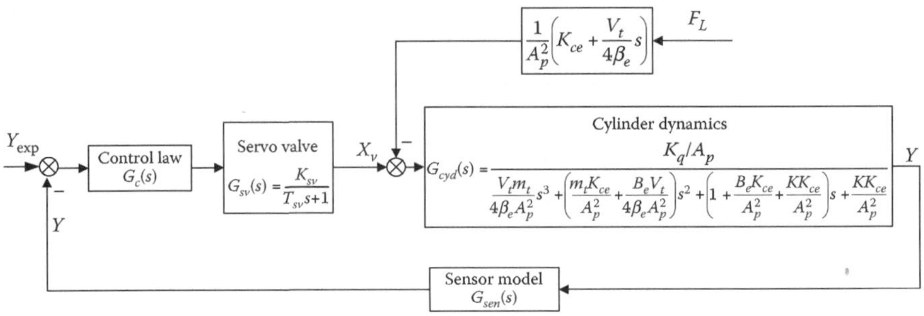

FIGURE 3.1
Linear feedback control loop of the electro-hydraulic system.

of the position sensor model is not temporarily considered, that is, it is a unit feedback, $G_{sen}(s) = 1$.

Consider the position demand Y_{exp} as the system input, and the actual cylinder position Y as the output. The open-loop transfer function of the position feedback control loop is given by

$$G_{ol}(s) = \frac{Y(s)}{u(s)} = G_c(s)G_{sv}(s)G_{cyd}(s)G_{sen}(s)$$
$$= \frac{\frac{K_q K_{sv}}{A_p(T_{sv}s+1)}}{\frac{V_t m}{4\beta_e A_p^2}s^3 + \left(\frac{m_t K_{ce}}{A_p^2} + \frac{B_e V_t}{4\beta_e A_p^2}\right)s^2 + \left(1 + \frac{bK_{ce}}{A_p^2} + \frac{V_t K}{4\beta_e A_p^2}\right)s + \frac{KK_{ce}}{A_p^2}}. \quad (3.1)$$

Remark 3.1

When the position feedback control loop is mainly considered, the external load F_L need not be substituted into Equation 3.1, which is different from Equation 2.20. In other words, F_L is compensated by the designed control u.

Now the effect of the external load F_L to the closed loop is analyzed. Consider F_L as the load disturbance input, and the actual cylinder position Y as the output. The closed loop from F_L to Y is shown as Figure 3.2.

The closed-loop transfer function from F_L to Y is given by

$$G_{yF_L}(s) = \frac{Y(s)}{F_L(s)}$$
$$= \frac{-\frac{1}{A_p^2}(K_{ce} + \frac{V_t}{4\beta_e}s)G_{cyd}(s)}{1 + G_c(s)G_{sv}(s)G_{cyd}(s)G_{sen}(s)}. \quad (3.2)$$

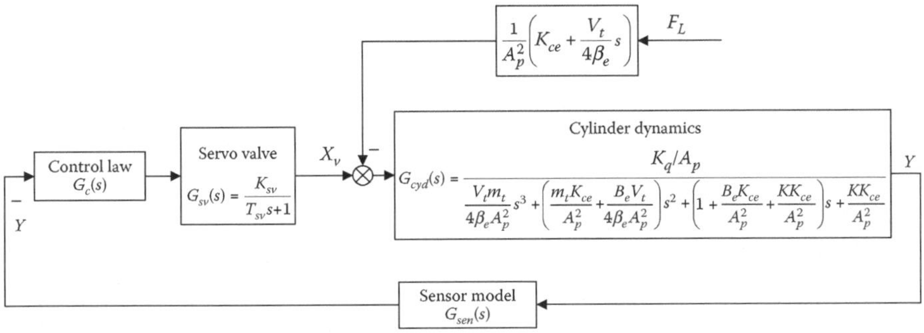

FIGURE 3.2
Closed loop from the load disturbance F_L to the cylinder position Y.

If the static gain of the controller $G_c(s)$ is obviously larger than 1, then Equation 3.2 can be simplified as follows:

$$G_{yF_L}(s) = \frac{-\frac{1}{A_p^2}K_{ce}}{G_c(s)G_{sv}(s)} = \frac{-\frac{1}{A_p^2}K_{ce}}{k_pK_{sv}}, \quad s \to 0 \Leftrightarrow t \to \infty, \tag{3.3}$$

where k_p is the static gain of $G_c(s)$, that is, $k_p = G_c(0)$.

From Equation 3.3, owing to the constants K_{ce} and K_{sv}, the negative effect of F_L to the system dynamic tracking performance is only declined by increasing the static gain of the controller $G_c(s)$. This denotes that a reasonable controller u can suppress the external disturbance F_L to a prescribed level.

Theorem 3.1

Consider the two transfer functions (3.1) and (3.2). There exists a linear PID controller $G_c(s)$ with sufficiently large static gain such that the closed-loop feedback control system as shown in Figure 3.1 has enough global stability margin and the external load disturbance $F_L(t)$ is suppressed into a sufficiently small finite boundary.

Proof. The position feedback control system is a linear system as shown in Figure 3.1. According to the global stability property of a linear control system, if there exists an isolated linear controller to guarantee the stability of the closed-loop feedback system, then the feedback control system is globally stable.

To address the open-loop transfer function of the position feedback control loop $G_{ol}(s)$, a suitable PID controller is designed as follows:

$$G_c(s) = k_p + \frac{k_i}{s} + k_d s, \tag{3.4}$$

where k_p, k_i, and k_d are the proportional, integral, and derivative control parameters, respectively.

If the three control parameters have no constraint, the closed-loop poles of $G_{ol}(s)$ can be arbitrarily configured to any position in the negative half plane. Thus, this PID controller is designed to guarantee the stability of the closed-loop feedback control system as shown in Figure 3.1. Especially, some reasonable PID parameters can guarantee that the open-loop transfer function $G_{ol}(s)$ has predefined amplitude margin, phase margin, and expected bandwidth as follows:

$$\begin{cases} A_m(\|G_{ol}(s)\|) > A_m^{\exp} \text{ dB} \\ P_m(\|G_{ol}(s)\|) > P_m^{\exp} \text{ deg} \\ \omega_c > \omega_c^{\exp} \end{cases}, \tag{3.5}$$

where $A_m^{\exp}$, $P_m^{\exp}$, and $\omega_c^{\exp}$ are the predefined performances of the relative stability margin.

On the other hand, from Equation 3.3, the sufficiently large static gain K_p of $G_c(s)$ can simultaneously reduce the disturbance effect caused by the external load $F_L(t)$ as follows:

$$\|y(\infty)\| = \frac{K_{ce}}{A_p^2 k_p K_{sv}} \|F_L(\infty)\| \leq \frac{K_{ce}}{A_p^2 k_p K_{sv}} F_L^{\max}, \tag{3.6}$$

where $F_L^{\max}$ is a finite boundary of $F_L(t)$. ■

However, in practice, the control variable u is constrained by the control saturation, that is, $u \leq u_{\max}$. Thus, the PID control law $G_c(s)$ should be appropriately designed to compromise the dynamic stable performance (3.5) and the disturbance suppression (3.6).

3.2 Mechatronics Plant Model Description

In this section, a two-DOF robotic arm is considered to be a typical mechatronics plant, which is driven by two electro-hydraulic actuators. The linear PID control of the electro-hydraulic system is designed to illustrate the motion control of the robotic arm.

3.2.1 Servo Valve Model Construction

Servo valve is the control element. It receives a voltage signal given by the controller and be converted to the spool position to control the oil flow in

TABLE 3.1

Servo Valve Parameters of Moog D633-R02K01M0NSM2

No.	Name	Symbol	Value	Unit
1	Response time	T_{sv}	12	ms
2	Rated flow	q_n	5	L/min
3	Rated pressure	p_n	70	bar
4	Maximal pressure	p_m	350	bar
5	Output current range	I_{ctrl}	4–20	mA
6	Control voltage range	U_{ctrl}	−10 to 10	V
7	Spool diameter	$D_{v\max}$	7.9	mm
8	Servo valve pressure loss	Δp_{vloss}	1.4	bar
9	Pipe pressure loss	Δp_{tube}	7.5	bar

the hydraulic cylinder. From Equation 2.8, the math model is simplified as follows:

$$G_{sv}(s) = \frac{x_v}{u} = \frac{K_{sv}}{T_{sv}s + 1}, \tag{3.7}$$

where x_v is the spool position of the servo valve, u is the control voltage of the servo valve, K_{sv} is the gain of the servo valve, and T_{sv} is the response time constant of the servo valve.

According to the production of Moog D633-R02K01M0NSM2, some performance parameters are shown in Table 3.1.

The maximal spool displacement is given by

$$x_{vm} = D_{vm}/2 = 3.95\,\text{mm}. \tag{3.8}$$

The gain of the servo valve K_{sv} is computed by

$$K_{sv} = \frac{x_{vm}}{U_{ctrl}} = \frac{3.95 \times 10^{-3}}{10} = 3.95 \times 10^{-4}\,\text{m/V}. \tag{3.9}$$

Thus, the first-order transfer function model of this servo valve is constructed as

$$\frac{x_v}{u} = \frac{K_{sv}}{T_{sv}s + 1} = \frac{3.95 \times 10^{-4}}{0.012s + 1}. \tag{3.10}$$

Owing to the servo valve hysteresis characteristics, the servo valve pressure loss is computed by

$$\Delta p_{vloss} = p_{s\max} \times \varsigma \times 2 = 350 \times 0.2\% \times 2 = 1.4\,\text{bar}, \tag{3.11}$$

where $p_{s\max}$ is the maximum pressure and ς is the parameter hysteresis characteristic.

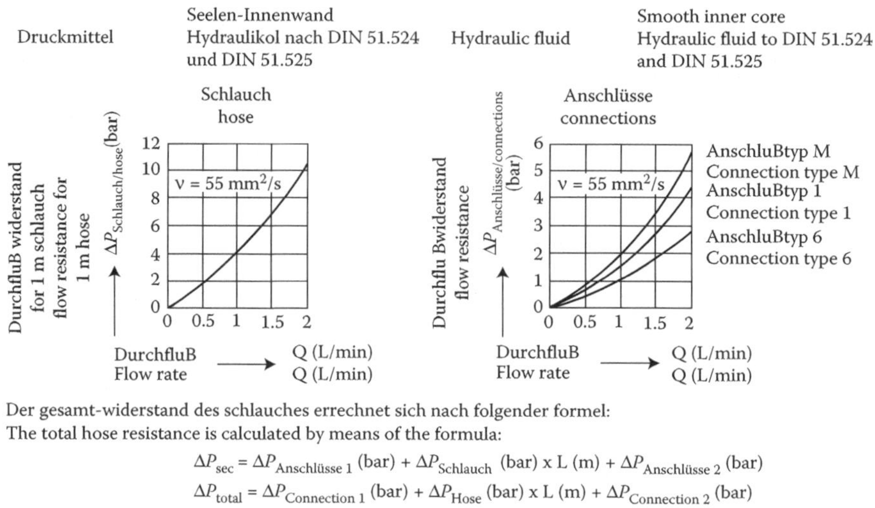

FIGURE 3.3
Change of pipe pressure loss Δp_{tube} from valve to cylinder.

According to the production of Hoerbiger-Hose-H3, the change of pipe pressure loss Δp_{tube} from valve to cylinder is shown in Figure 3.3.

Ignoring the pressure loss of the connector, the pipe pressure loss from the proportional valve outlet to the inlet of the hydraulic cylinder is computed by

$$\Delta p_{tube} = k_{\Delta p/Q} \times Q_a \times l_{tube}, \tag{3.12}$$

where $k_{\Delta p/Q} = 5$ is the approximate slope of the pipe pressure loss with respect to flow, $Q_a = 3$ L/min is the average flow of this hydraulic system, and $l_{tube} = 0.5$ m is the pipe length from valve to cylinder. Thus, Δp_{tube} is computed approximately by

$$\Delta p_{tube} = k_{\Delta p/Q} \times Q_{\max} \times l_{tube} = 5 \times 2.98 \times 0.5 = 7.45\,\text{bar}. \tag{3.13}$$

If the maximal output pressure from the pump is the supply pressure p_s, the maximal no-load flow of the servo valve is computed by

$$q_{0m} = q_n\sqrt{\frac{p_s}{p_n}} = 5 \times \sqrt{\frac{p_s - \Delta p_{vloss} - \Delta p_{tube}}{70}}\ \text{L/min}. \tag{3.14}$$

Thus, the no-load flow gain of the proportional valve is computed by

$$K_{q0} = \frac{q_{0m}}{x_{vm}}\ \text{m}^2/\text{s}. \tag{3.15}$$

Then the load flow gain of the servo valve is computed by

$$K_q = K_{q0}\sqrt{\frac{p_s - p_l - \Delta p_{vloss} - \Delta p_{tube}}{p_s}}, \tag{3.16}$$

where p_l is the load pressure of the hydraulic cylinder, which is estimated by the second-order Lagrangian math model.

3.2.2 Hydraulic Cylinder Model

The hydraulic cylinder is the actuator of the control loop to output the position and pressure. In this example, the asymmetric hydraulic cylinder is adopted as shown in Figure 2.1b. Some parameters of the hydraulic cylinder "Hoerbiger LB6-1610-0080-4M" are shown in Table 3.2.

The cross-section of the rod chamber is computed by

$$A_p = \frac{\pi D_c^2}{4} = \frac{\pi \times (16 \times 10^{-3})^2}{4} = 2.01 \times 10^{-4}\,\text{m}^2. \tag{3.17}$$

The total volume of the cylinder is as follows:

$$V_t = A_p L_c = 1.83 \times 10^{-5}\,\text{m}^3. \tag{3.18}$$

The pressure gain of the valve K_p is computed by

$$K_p = \frac{p_s - \Delta p_{vloss} - \Delta p_{tube}}{x_{vm}}. \tag{3.19}$$

The flow-pressure coefficient of the valve K_c and the coefficient K_{ce} are computed by

$$\begin{aligned} K_c &= K_q/K_p, \\ K_{ce} &= K_c + C_{tl}, \end{aligned} \tag{3.20}$$

TABLE 3.2

Parameters of the Hydraulic Cylinder

No.	Name	Symbol	Value	Unit
1	Cylinder length	L_c	91	mm
2	Piston stroke	$d_{r\max}$	79	mm
3	Cylinder diameter	D_c	16	mm
4	Rod diameter	D_{rod}	10	mm
5	Load mass of No. 2 cylinder	m_{2f}	1.778	kg
6	Load mass of No. 1 cylinder	m_{1f}	3.550	kg
7	Load spring stiffness	K	0	N/m
8	Load damping	B_e	50	Ns/m
9	Elastic modulus	β_e	7000	bar

where C_{tl} is the total leakage coefficient of the cylinder.

From the linearized model of the single-rod electro-hydraulic actuator (2.24), the load flow gain of the valve K_q and the flow-pressure coefficient of the valve K_c are rewritten as follows:

$$\begin{cases} K_{qa} = K_q \times \sqrt{\dfrac{2}{1+\upsilon^3}} \\ K_{qb} = K_q \times \sqrt{\dfrac{2(\upsilon(p_s - \Delta p_{vloss} - \Delta p_{tube}) + p_l)}{(p_s - \Delta p_{vloss} - \Delta p_{tube} - p_l)(1+\upsilon^3)}} \end{cases}, \tag{3.21}$$

$$\begin{cases} K_{ca} = K_c \times \sqrt{\dfrac{2}{1+\upsilon^3}} \\ K_{cb} = K_c \times \sqrt{\dfrac{2(p_s - \Delta p_{vloss} - \Delta p_{tube} - p_l)}{(\upsilon(p_s - \Delta p_{vloss} - \Delta p_{tube}) + p_l)(1+\upsilon^3)}} \end{cases}, \tag{3.22}$$

where $\upsilon \leq 1$ is the annulus area ratio of two chambers, and the coefficient K_{ce} is rewritten as $K_{cea} = K_{ca} + C_{tl}$, $K_{ceb} = K_{cb} + C_{tl}$.

3.2.3 Mechanical Motion Model

The motion control mechanism of the two-DOF robotic arm is shown in Figure 3.4 [51]. The shoulder and elbow joints are driven by two single-rod electro-hydraulic actuators.

At first, the general framework for the mechanical movement must be given in Figure 3.5. Some mechanical parameters of the two-DOF robotic arm are shown in Table 3.3.

The initial load mass is computed by

$$m_{f0} = m_{2f0} - m_2 = 0.808 - 0.739 = 0.069\,\text{kg}. \tag{3.23}$$

If the load mass is m_f, the moment of inertia of the load rotation from the elbow is given by

$$I_{fp2} = I_{2f} - I_2 - m_2(x_c - P_2P_{m20})^2 + m_f(P_2P_3 - x_c)^2. \tag{3.24}$$

If the load mass is m_{f0}, the centroid distance from the elbow to the forearm (including initial load) x_{c0} is P_2P_{m2}. So the moment of inertia of the initial load rotation from the elbow is

$$\begin{aligned} I_{fp20} &= 0.0218 - 0.0145 - 0.730 \times (0.122 - 0.103)^2 + 0.069 \times (0.33 - 0.122)^2 \\ &= 0.010018\,\text{kgm}^2. \end{aligned} \tag{3.25}$$

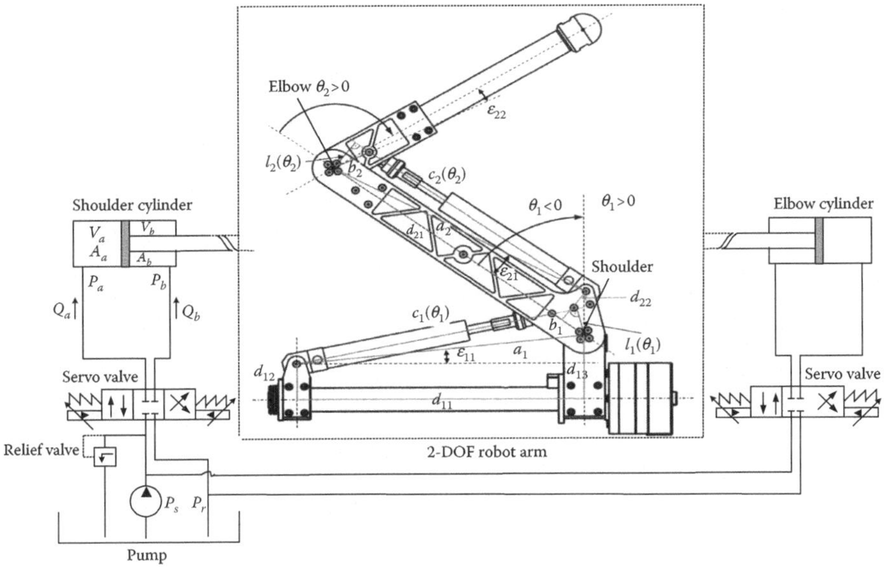

FIGURE 3.4
Motion control mechanism of the two-DOF robotic arm.

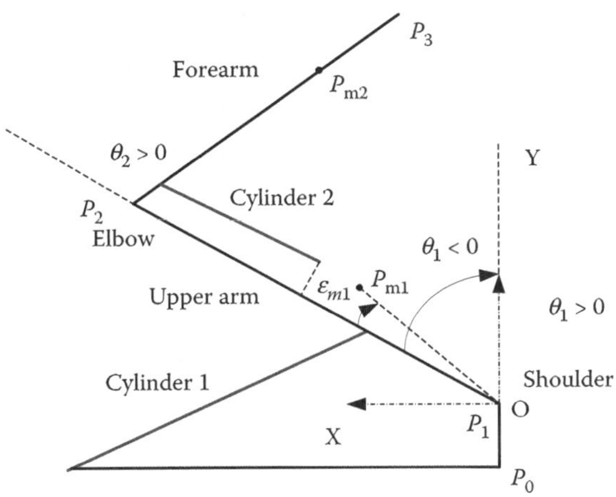

FIGURE 3.5
General framework for the mechanical movement.

If the load is a spherical particle, the moment of inertia of the load rotation from the elbow is proportional to the mass of the load. So it is computed by

$$I_{fp2} = \frac{m_f}{m_{f0}} I_{fp20} = \frac{1.039}{0.069} \times 0.010018 = 0.1508573\,\text{kgm}^2. \qquad (3.26)$$

TABLE 3.3

Mechanical Parameters of the Two-DOF Robotic Arm

No.	Name	Symbol	Value	Unit
1	Forearm mass	m_2	0.739	kg
2	Upper arm mass with cylinder	m_1	1.772	kg
3	Load mass	m_f	1.039	kg
4	Moment of inertia for forearm around the elbow	I_2	0.0145	kgm^2
5	Moment of inertia for upper arm around the shoulder	I_1	0.0713	kgm^2
6	Forearm rotation range	θ_2	20–140	deg
7	Upper arm rotation range	θ_1	−70 to 50	deg
8	Forearm around the elbow inertia with load	I_{2f}	0.0218	kgm^2
9	Forearm mass with load	m_{2f0}	0.808	kg
10	Upper arm length	P_1P_2	0.35	m
11	Distance from centroid upper arm to shoulder	P_1P_{m1}	0.164	m
12	Forearm length	P_2P_3	0.33	m
13	Distance from centroid forearm to elbow	P_2P_{m2}	0.122	m
14	Distance from centroid forearm to elbow without load	P_2P_{m20}	0.103	m

When the current load is 1.039 kg, the distance from the centroid forearm to the elbow (including the current load) is computed by

$$x_c = \frac{m_f \times P_2P_3 + m_2 \times P_2P_{m20}}{m_f + m_2} = 0.23565\,\text{m}. \tag{3.27}$$

Substituting I_{fp2} and x_c into the above equation, the equivalent moment of inertia of the forearm with the new load is computed by

$$I'_{2f} = 0.16911\,\text{kgm}^2. \tag{3.28}$$

The equivalent moment of inertia of the forearm with the new load rotated at its centroid is computed by

$$I'_{2fm} = I'_{2f} - (m_2 + m_f)x_c^2 = 0.07038\,\text{kgm}^2. \tag{3.29}$$

The equivalent moment of inertia of the upper arm with the cylinder rotated at its centroid is computed by

$$I_{1m} = I_1 - m_1(P_1P_{m1})^2 = 0.02364\,\text{kgm}^2. \tag{3.30}$$

According to Lagrange's theorem, a two-link dynamic model is established. The Lagrange function is defined as follows:

$$L = KE - V, \tag{3.31}$$

where *KE* is the kinetic energy of the system and *V* is the potential.

Thus, the Lagrange equation of the two-DOF robotic arm is given by

$$\boldsymbol{T} = \frac{\partial}{\partial t}\frac{\partial L}{\partial \dot{\boldsymbol{q}}} - \frac{\partial L}{\partial \boldsymbol{q}}, \tag{3.32}$$

where $\boldsymbol{T}$ is the generalized force vectors including force and torque and $\boldsymbol{q}$ is the generalized coordinates vector of the system.

The coordinate X-O-Y is constructed in Figure 3.5. The shoulder joint P_1 is the origin point. A hinge point P_1 is fixed with the bedframe. The position vector of P_1P_{m1} is as follows:

$$\begin{aligned} \boldsymbol{r}_{oPm1/o} &= \left[P_1P_{m1}\sin(-\theta_1 - \varepsilon_{m1}) \quad P_1P_{m1}\cos(-\theta_1 - \varepsilon_{m1})\right]^T \\ &= \left[-P_1P_{m1}\sin(\theta_1 + \varepsilon_{m1}) \quad P_1P_{m1}\cos(\theta_1 + \varepsilon_{m1})\right]^T. \end{aligned} \tag{3.33}$$

The velocity vector of P_{m1} is computed by

$$\boldsymbol{v}_{Pm1/o} = \left[-P_1P_{m1}\dot{\theta}_1\cos(\theta_1 + \varepsilon_{m1}) \quad -P_1P_{m1}\dot{\theta}_1\sin(\theta_1 + \varepsilon_{m1})\right]^T. \tag{3.34}$$

The position vector of P_1P_2 is given by

$$\begin{aligned} r_{oP2/o} &= \left[P_1P_2\sin(-\theta_1 - \varepsilon_{m1}) \quad P_1P_2\cos(-\theta_1 - \varepsilon_{m1})\right]^T \\ &= \left[-P_1P_2\sin(\theta_1 + \varepsilon_{m1}) \quad P_1P_2\cos(\theta_1 + \varepsilon_{m1})\right]^T. \end{aligned} \tag{3.35}$$

The velocity vector of P_2 is computed by

$$\boldsymbol{v}_{P2/o} = \left[-P_1P_2\dot{\theta}_1\cos(\theta_1 + \varepsilon_{m1}) \quad -P_1P_2\dot{\theta}_1\sin(\theta_1 + \varepsilon_{m1})\right]^T. \tag{3.36}$$

The position vector of P_{m2} is given by

$$\begin{aligned} \boldsymbol{r}_{oPm2/o} &= r_{oP2/o} + [-P_2P_{m2}\sin(\theta_1 + \theta_2), P_2P_{m2}\cos(\theta_1 + \theta_2)]^T \\ &= [-P_1P_2\sin(\theta_1 + \varepsilon_{m1}) - P_2P_{m2}\sin(\theta_1 + \theta_2), \\ &\quad P_1P_2\cos(\theta_1 + \varepsilon_{m1}) + P_2P_{m2}\cos(\theta_1 + \theta_2)]^T. \end{aligned} \tag{3.37}$$

The velocity vector of P_{m2} is given by

$$\boldsymbol{v}_{Pm2/o} = \begin{bmatrix} -P_1P_2\dot{\theta}_1\cos(\theta_1 + \varepsilon_{m1}) - P_2P_{m2}(\dot{\theta}_1 + \dot{\theta}_2)\cos(\theta_1 + \theta_2) \\ -P_1P_2\dot{\theta}_1\sin(\theta_1 + \varepsilon_{m1}) - P_2P_{m2}(\dot{\theta}_1 + \dot{\theta}_2)\sin(\theta_1 + \theta_2) \end{bmatrix}. \tag{3.38}$$

Then, the kinetic energy of the upper arm including cylinder 2 is computed by

$$KE_{pm1} = 0.5m_1 v_{Pm1/o} \cdot v_{Pm1/o} + 0.5I_1\dot{\theta}_1 \cdot \dot{\theta}_1. \tag{3.39}$$

The kinetic energy of the forearm including the load is computed by

$$KE_{pm2} = 0.5m_{2f} v_{Pm2/o} \cdot v_{Pm2/o} + 0.5I_{2f}(\dot{\theta}_1 + \dot{\theta}_2) \cdot (\dot{\theta}_1 + \dot{\theta}_2). \tag{3.40}$$

The gravitational potential energy of the upper arm including cylinder 2 is computed by

$$V_{pm1} = m_1 g P_1 P_{m1} \cos(\theta_1 + \varepsilon_{m1}). \tag{3.41}$$

The gravitational potential energy of the forearm including the load is computed by

$$V_{pm2} = m_{2f} g(P_1P_2 \cos(\theta_1 + \varepsilon_{m1}) + P_2P_{m2} \cos(\theta_1 + \theta_2)). \tag{3.42}$$

So, the total energy of the two-link system is computed by

$$\begin{aligned}
L &= KE_{pm1} + KE_{pm2} - V_{pm1} - V_{pm2} \\
&= 0.5m_1 P_1P_{m1}^2 \dot{\theta}_1^2 + 0.5I_1\dot{\theta}_1^2 + 0.5m_{2f}[P_1P_2^2\dot{\theta}_1^2 + P_2P_{m2}^2(\dot{\theta}_1^2 + \dot{\theta}_2^2) \\
&\quad + 2P_1P_2 \cdot P_2P_{m2}\dot{\theta}_1(\dot{\theta}_1 + \dot{\theta}_2)\cos(\theta_2 - \varepsilon_{m1})] + 0.5I_{2f}(\dot{\theta}_1 + \dot{\theta}_2)^2 \\
&\quad - m_1 g P_1P_{m1} \cos(\theta_1 + \varepsilon_{m1}) \\
&\quad - m_{2f} g[P_1P_2 \cos(\theta_1 + \varepsilon_{m1}) + P_2P_{m2} \cos(\theta_1 + \theta_2)].
\end{aligned} \tag{3.43}$$

Using the Lagrange equation, if the angle vector $\boldsymbol{q} = [\theta_1, \theta_2]^T$, then the kinetic equation of the two-link system is given as follows:

$$\boldsymbol{H}(\boldsymbol{q})\ddot{\boldsymbol{q}} + \boldsymbol{C}(\boldsymbol{q}, \dot{\boldsymbol{q}})\dot{\boldsymbol{q}} + \boldsymbol{G}(\boldsymbol{q}) = \boldsymbol{T}, \tag{3.44}$$

where

$$\boldsymbol{H} = \begin{bmatrix} H_{11} & H_{12} \\ H_{21} & H_{22} \end{bmatrix}, \quad \boldsymbol{C} = \begin{bmatrix} C_{11} & C_{12} \\ C_{21} & C_{22} \end{bmatrix}, \quad \boldsymbol{G} = \begin{bmatrix} G_1 \\ G_2 \end{bmatrix}, \quad \boldsymbol{T} = \begin{bmatrix} T_u \\ T_f \end{bmatrix}, \tag{3.45}$$

$$
\begin{aligned}
H_{11} &= I_1 + I_{2f} + m_1 P_1 P_{m1}^2 + m_{2f} P_1 P_2^2 + m_{2f} P_2 P_{m2}^2 \\
&\quad + 2m_{2f} P_1 P_2 \cdot P_2 P_{m2} \cos(\theta_2 - \varepsilon_{m1}), \\
H_{12} &= I_{2f} + m_{2f} P_1 P_2 \cdot P_2 P_{m2} \cos(\theta_2 - \varepsilon_{m1}), \\
H_{22} &= I_{2f} + m_{2f} P_2 P_{m2}^2, \\
C_{11} &= -2m_{2f} P_1 P_2 \cdot P_2 P_{m2} \dot{\theta}_2 \sin(\theta_2 - \varepsilon_{m1}), \\
C_{12} &= -m_{2f} P_1 P_2 \cdot P_2 P_{m2} \dot{\theta}_2 \sin(\theta_2 - \varepsilon_{m1}), \\
C_{21} &= m_{2f} P_1 P_2 \cdot P_2 P_{m2} \dot{\theta}_1 \sin(\theta_2 - \varepsilon_{m1}), \\
C_{22} &= 0, \\
G_1 &= -m_1 g P_1 P_{m1} \sin(\theta_1 + \varepsilon_{m1}) \\
&\quad - m_{2f} g [P_1 P_2 \sin(\theta_1 + \varepsilon_{m1}) + P_2 P_{m2} \sin(\theta_1 + \theta_2)], \\
G_2 &= -m_{2f} g P_2 P_{m2} \sin(\theta_1 + \theta_2).
\end{aligned}
\tag{3.46}
$$

After the kinetic model is constructed, according to the changes of the input commands, a rate, and secondary rate of commands, two driving torques of a two-link system are computed.

Then, two cylinder dynamic lengths are computed by

$$
\begin{cases}
c_1(\theta_1) = \sqrt{a_1^2 + b_1^2 - 2a_1 b_1 \cos(\pi/2 + \theta_1 + \varepsilon_{11})} \\
c_2(\theta_2) = \sqrt{a_2^2 + b_2^2 - 2a_2 b_2 \cos(\pi - \theta_2 - \varepsilon_{21} - \varepsilon_{22})}
\end{cases},
\tag{3.47}
$$

where $a_1 = a_2 = 0.3219\,\mathrm{m}$, $b_1 = b_2 = 0.045\,\mathrm{m}$, $\varepsilon_{11} = 6.24°$, $\varepsilon_{21} = 8.04°$, and $\varepsilon_{22} = 6°$.

Thus, the two dynamic force arms are given by

$$
\begin{cases}
l_1(\theta_1) = a_1 \sin\left(\arccos\left(\dfrac{a_1^2 + c_1(\theta_1)^2 - b_1^2}{2a_1 c_1(\theta_1)}\right)\right) \\
l_2(\theta_2) = a_2 \sin\left(\arccos\left(\dfrac{a_2^2 + c_2(\theta_2)^2 - b_2^2}{2a_2 c_2(\theta_2)}\right)\right)
\end{cases},
\tag{3.48}
$$

where the ranges of two joint angles are $-70° \le \theta_1 \le 50°$, $20° \le \theta_2 \le 140°$.

In this robotic plant, the simulation results of two cylinder dynamic lengths and force arms are as shown in Figure 3.6

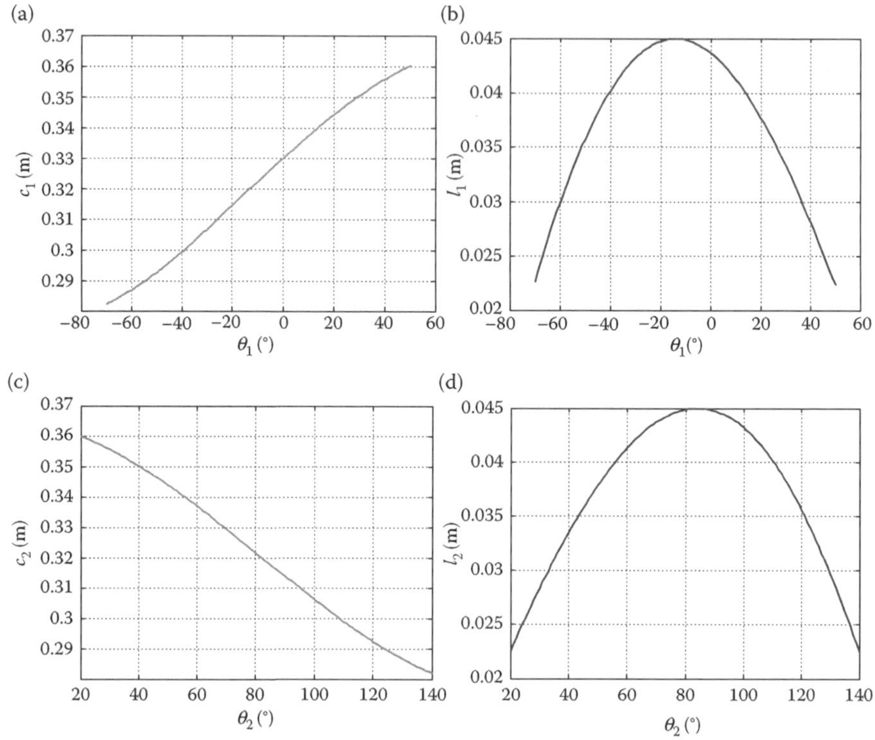

FIGURE 3.6
Two cylinder dynamic lengths and force arms. (a) The dynamic length of the upper arm cylinder c_1. (b) The dynamic force arm of the upper arm cylinder l_1. (c) The dynamic length of the forearm cylinder c_2. (d) The dynamic force arm of the forearm cylinder l_2.

3.3 System Performance Analysis

The system performance should be considered due to the selected motor and the pump. From Equation 2.12, the dynamic loads F_L of the robotic arm depend on the hydraulic cylinder position y, velocity $\dot{y}$, and acceleration $\ddot{y}$. Thus, this system performance includes motor performance, output pressure analysis of the pump, and dynamic pressure of the cylinder with load.

3.3.1 Motor Performance

The maximal allowable speed of the pump is 3000 rpm. According to the mechanical efficiency curve in the motor type "BSM63N-375," the continuous rated torque is 1.85 Nm as the motor is operated in 3000 rpm with load

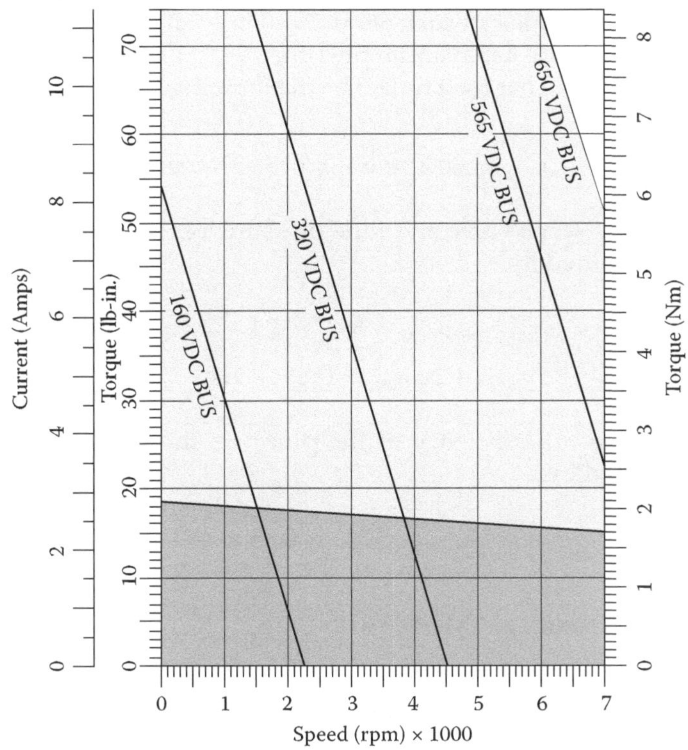

FIGURE 3.7
Mechanical properties curve of the motor with load.

as shown in Figure 3.7. So, at this operated point, the motor rated power is

$$P_m = \frac{1.85 \times 3000 \times 2 \times \pi}{60} = 581\,\text{W}. \tag{3.49}$$

3.3.2 Output Pressure Analysis of Pump

The fixed pump displacement $\nu = 3.14\,\text{mL/r}$. The continuous torque of motor $T_c = 2.09\,\text{Nm}$. The peak torque is $T_{peak} = 8.36\,\text{Nm}$. So, the rated pressure and the maximum pressure of the pump is computed by

$$\begin{aligned} P_n &= \frac{T_c \times 2 \times \pi \times 10}{\nu} \times \eta = 39.7\,\text{bar}, \\ P_{\max} &= \frac{T_{peak} \times 2 \times \pi \times 10}{\nu} \times \eta = 158.9\,\text{bar}. \end{aligned} \tag{3.50}$$

where $\eta = 0.95$ is the volumetric efficiency coefficient of the pump.

The peak motor torque can stick about 2 s. This indicates only a short time for the pump to provide a maximum pressure 158.9 bar. So the operated pressure is no more than 40 bar long time. The maximum output flow of the pump is computed by

$$Q_{\max} = 3000 \times \nu \times \eta = 8.95\,\mathrm{L/min}. \tag{3.51}$$

Considering the servo valve and pipe pressure loss, the actual pressure of cylinders is computed by

$$\begin{aligned} P_{an} &= P_n - \Delta p_{vloss} - \Delta p_{tube} = 39.7 - 1.4 - 7.5 = 30.8\,\mathrm{bar}, \\ P_{a\max} &= P_{\max} - \Delta p_{vloss} - \Delta p_{tube} = 158.9 - 1.4 - 7.5 = 150\,\mathrm{bar}. \end{aligned} \tag{3.52}$$

Considering the full efficiency of the pump ς, the output power of the pump is computed by

$$P_{out} = P_m \times \varsigma = 581 \times 0.85 = 494\,\mathrm{W}. \tag{3.53}$$

3.3.3 Dynamic Pressure of Cylinder with Load

The pressure of the cylinder with load is a dynamic value, which depends on the mass of the load, joint rotational movement, angular velocity, and angular acceleration. The motion model of a two-link mechanical structure with load is built by MATLAB®/Simulink® shown in Figure 3.8. Two joint angles θ_1, θ_2 are imported like these sinusoidal variation.

$$\begin{cases} \theta(t) = \dfrac{\theta_{\min} + \theta_{\max}}{2} + \dfrac{\theta_{\max} - \theta_{\min}}{2} \sin(2\pi f t) \\ \dot{\theta}(t) = \dfrac{\theta_{\max} - \theta_{\min}}{2} \times 2\pi \times \cos(2\pi f t) \\ \ddot{\theta}(t) = -\dfrac{\theta_{\max} - \theta_{\min}}{2} \times 4\pi^2 \times \sin(2\pi f t) \end{cases}. \tag{3.54}$$

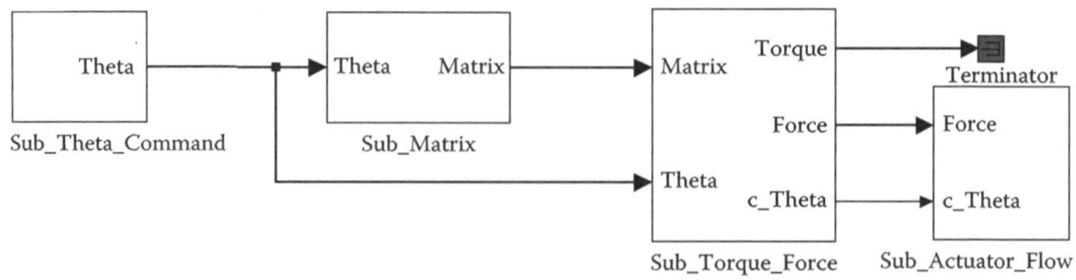

FIGURE 3.8
Simulation model for solvering the dynamic pressure of the cylinder with load.

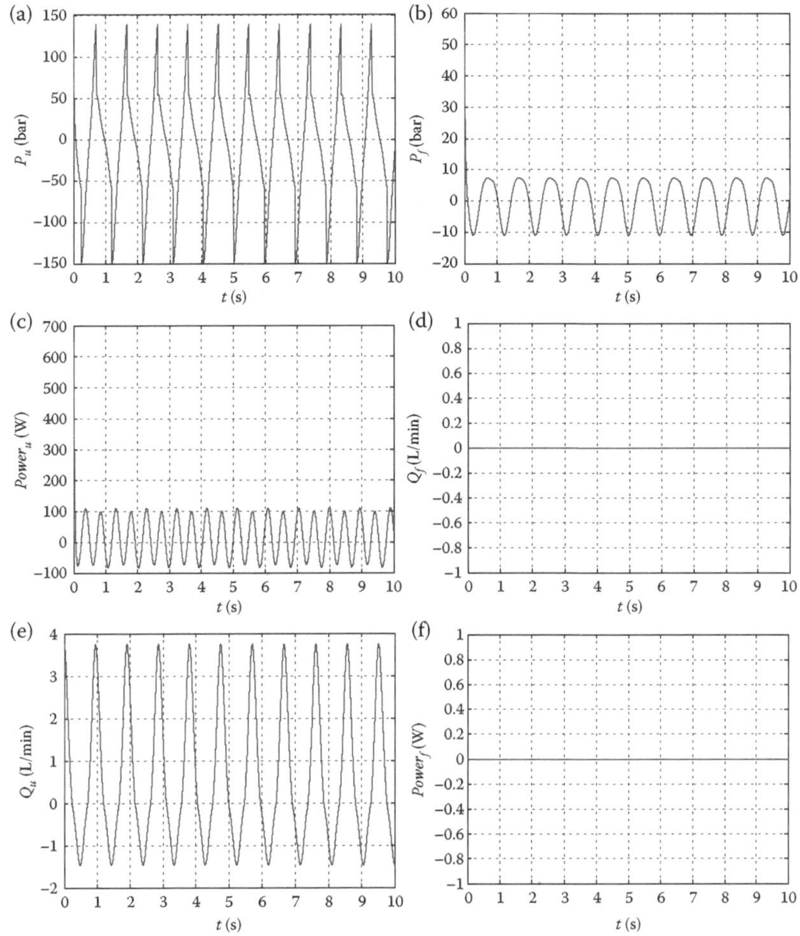

FIGURE 3.9
Simulation result with the elbow joint $\theta_2 = 140°$, the sinusoidal input of the shoulder angle $f_u = 1.05\,\text{Hz}$. (a) The pressure of the upper arm cylinder with load p_u. (b) The pressure of the forearm cylinder with load p_f. (c) The flow of the upper arm cylinder with load Q_u. (d) The flow of the forearm cylinder with load Q_f. (e) The power consumption of the upper arm cylinder with load $Power_u$. (f) The power consumption of the forearm cylinder with load $Power_f$.

Case 1: The pressure of the cylinder supporting the shoulder

If the elbow joint is fixed, $\theta_2 = \theta_{2\,\text{min}} = 140°$ and the load $m_f = 1.039\,\text{kg}$, and the shoulder angle is sinusoidal input, the simulation result is acquired as shown in Figure 3.9.

As shown in Figure 3.9, if the condition of the frequency of θ_1 is 1.05 Hz, the flow of the upper arm cylinder with load is 150 bar. It reaches the maximum motor torque value. But the power of the upper arm cylinder is 110 W, which is less than the rated power 494 W. At this moment, the flow of the upper

TABLE 3.4

Parametric Analysis of the Shoulder Hydraulic Actuator

Condition	Maximum Frequency	Maximum Pressure	Maximum Flow	Maximum Power
$\theta_2 = 140°$	1.05 Hz	150 bar	3.75 L/min	110 W
$\theta_2 = 20°$	0.7 Hz	– 149 bar	2.77 L/min	76 W

arm cylinder is 3.75 L/min, which is less than the output flow of the pump 8.95 L/min. These results indicate that the frequency of the shoulder angle should not exceed 1.05 Hz. The parametric analysis of the shoulder hydraulic actuator is shown in Table 3.4.

Case 2: The pressure of the cylinder supporting the elbow
If the shoulder joint is fixed, $\theta_1 = \theta_{1\,\min} = -70°$ and the load $m_f = 1.039$ kg, and the elbow angle is sinusoidal input, the simulation result is acquired as shown in Figure 3.10.

As shown in Figure 3.10, if the condition of the frequency of θ_2 is 1.85 Hz, the flow of the forearm cylinder with load is 150 bar. It reaches the maximum torque value. But the power of the forearm cylinder is 226 W, which is less than the rated power 494 W. At this moment, the flow of the forearm cylinder is 6.6 L/min, which is less than the output flow of the pump 8.95 L/min. These results indicate that the frequency of the elbow angle should not exceed 1.85 Hz. The parametric analysis of the elbow hydraulic actuator is shown in Table 3.5.

Case 3: The pressure of the cylinder with two joints in simultaneously motion
If two joints change simultaneously and the load $m_f = 1.039$ kg, the simulation results are acquired as shown in Figure 3.11.

As shown in these figures, if the condition of the frequency of θ_1 is 0.6 Hz and θ_2 is 1 Hz, the flow of the upper arm cylinder with load is 150 bar. It reaches the maximum torque value. But the power of the upper arm cylinder is 70 W, which is less than the rated power 494 W. Also, the power of the forearm cylinder is only 47 W. At this moment, the total flow of two cylinders is 5 L/min, which is less than the output flow of the pump 8.95 L/min. These results indicate that the frequencies of shoulder and elbow angles should not exceed 0.6 and 1 Hz respectively. The parametric analysis of two hydraulic actuators is shown in Table 3.6.

3.4 PID Controller Design

The actual angles of two joints are measured by encoders. The errors between actual angles and command angles are as input of the control system. The controller inputs the control voltage into proportional valves and the spool dynamic positions are regulated to drive the cylinder motion.

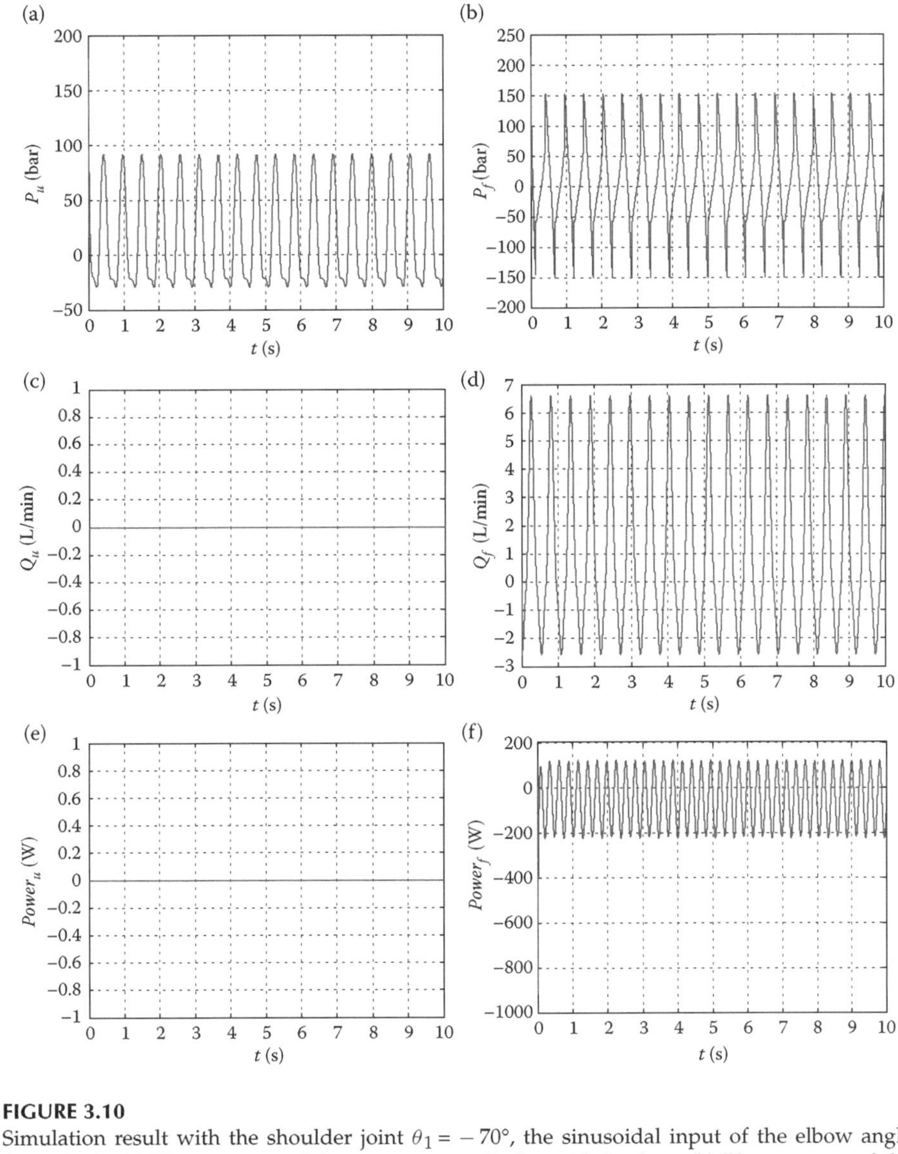

FIGURE 3.10
Simulation result with the shoulder joint $\theta_1 = -70°$, the sinusoidal input of the elbow angle $f_f = 1.85$ Hz. (a) The pressure of the upper arm cylinder with load p_u. (b) The pressure of the forearm cylinder with load p_f. (c) The flow of the upper arm cylinder with load Q_u. (d) The flow of the forearm cylinder with load Q_f. (e) The power consumption of the upper arm cylinder with load $Power_u$. (f) The power consumption of the forearm cylinder with load $Power_f$.

Considering the dynamic external load on two hydraulic actuators, the supply pressure is chosen as follows:

$$p_s = \frac{3}{2} p_{l\max}. \tag{3.55}$$

TABLE 3.5

Parametric Analysis of Elbow Hydraulic Actuator

Condition	Maximum Frequency	Maximum Pressure	Maximum Flow	Maximum Power
$\theta_1 = -70°$	1.85 Hz	150 bar	6.6 L/min	226 W
$\theta_1 = 50°$	1.95 Hz	– 150 bar	7.0 L/min	255 W

The supply pressure p_s is 150 bar. $p_{l\,\max}$ is the maximal pressure of the cylinder with load. If the load mass is 3 kg, the value is 100 bar of the upper arm cylinder, which satisfies this relation. So the load mass is chosen as 3 kg sufficiently. Owing to the asymmetrical hydraulic cylinder, the open-loop transfer function of this system is shown as follows:

1. If the hydraulic cylinder is extended, that is, $\dot{y} \geq 0$, then

$$G_{ol} = \frac{G_c(s)\dfrac{K_{qa}}{A_a}\dfrac{K_{sv}}{T_{sv}s+1}}{\dfrac{V_t m_t}{2(1+\upsilon^2)\beta_e A_a^2}s^3 + \left(\dfrac{m_t K_{cea}}{A_a^2} + \dfrac{B_e V_t}{2(1+\upsilon^2)\beta_e A_a^2}\right)s^2 + \left(1 + \dfrac{B_e K_{cea}}{A_a^2} + \dfrac{V_t K}{2(1+\upsilon^2)\beta_e A_a^2}\right)s + \dfrac{KK_{cea}}{A_a^2}}. \quad (3.56)$$

2. If the hydraulic cylinder is retracted, that is, $\dot{y} < 0$, then

$$G_{ol} = \frac{G_c(s)\dfrac{K_{qb}}{A_a}\dfrac{K_{sv}}{T_{sv}s+1}}{\dfrac{V_t m_t}{2(1+\upsilon^2)\beta_e A_a^2}s^3 + \left(\dfrac{m_t K_{ceb}}{A_a^2} + \dfrac{B_e V_t}{2(1+\upsilon^2)\beta_e A_a^2}\right)s^2 + \left(1 + \dfrac{B_e K_{ceb}}{A_a^2} + \dfrac{V_t K}{2(1+\upsilon^2)\beta_e A_a^2}\right)s + \dfrac{KK_{ceb}}{A_a^2}}, \quad (3.57)$$

where $G_c(s)$ is the PID control law.

The maximal supply pressure p_s is 150 bar and the maximal load mass m_f is 3 kg. The control law is designed by MATLAB/SISO tool like this

$$G_c(s) = \left(k_p + \frac{k_i}{s}\right) \times \frac{1}{1+b_1 s}, \quad (3.58)$$

where k_p and k_i are the proportional and integral control parameters, and b_1 is the lag parameter. This hysteresis element $1/(1+b_1 s)$ can improve the dynamic quality of the control variable u and suppress the control saturation as the initial large error existing in the system.

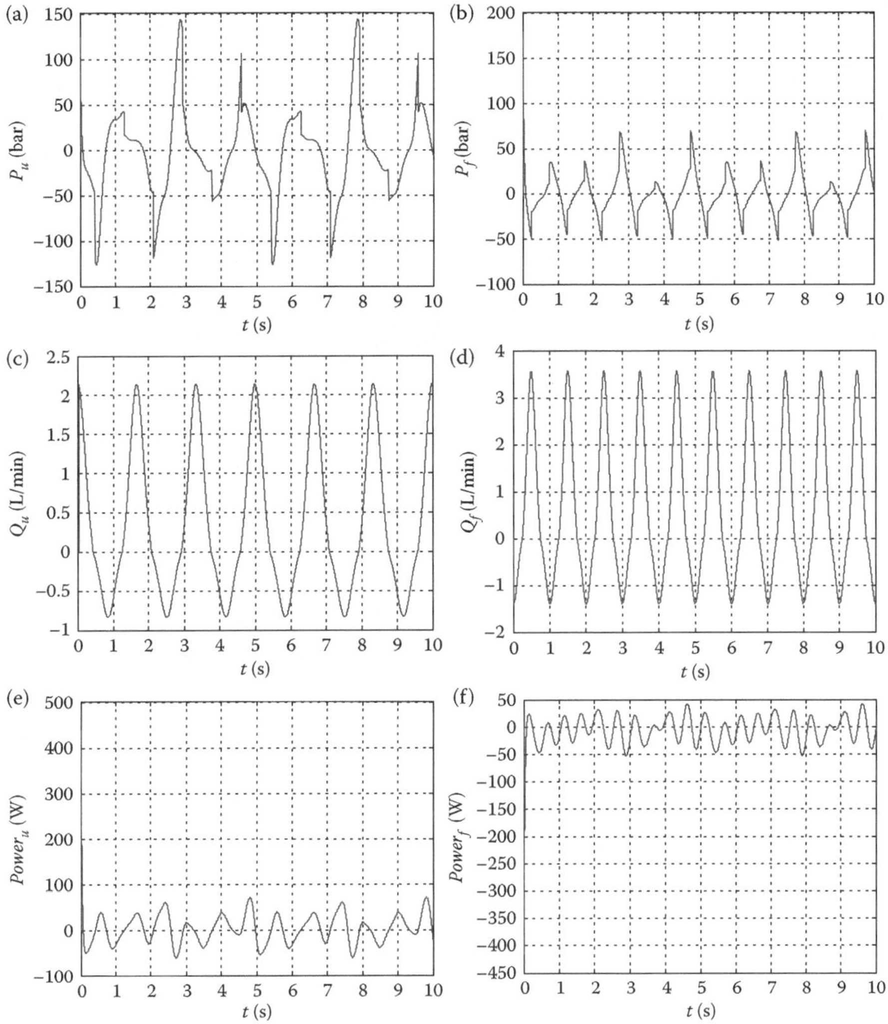

FIGURE 3.11
Simulation result with two sinusoidal inputs of the shoulder angle $f_u = 0.6$ Hz and the elbow angle $f_f = 1$ Hz. (a) The pressure of the upper arm cylinder with load p_u. (b) The pressure of the forearm cylinder with load p_f. (c) The flow of the upper arm cylinder with load Q_u. (d) The flow of the forearm cylinder with load Q_f. (e) The power consumption of the upper arm cylinder with load $Power_u$. (f) The power consumption of the forearm cylinder with load $Power_f$.

The designed results of control design for the upper arm cylinder is shown as follows:

1. If the hydraulic cylinder is extended, that is, $\dot{y} \geq 0$, the control parameters and the lag parameter are $k_p = 178.5$, $k_i = 21$, and $b_1 = 0.073$.

TABLE 3.6

Parametric Analysis of Two Joint Angles Simultaneously Motion

Joint	Maximum Frequency	Maximum Pressure	Maximum Flow	Maximum Power
Shoulder	0.6 Hz	145 bar	2.14 L/min	70 W
Elbow	1 Hz	70 bar	3.6 L/min	47 W

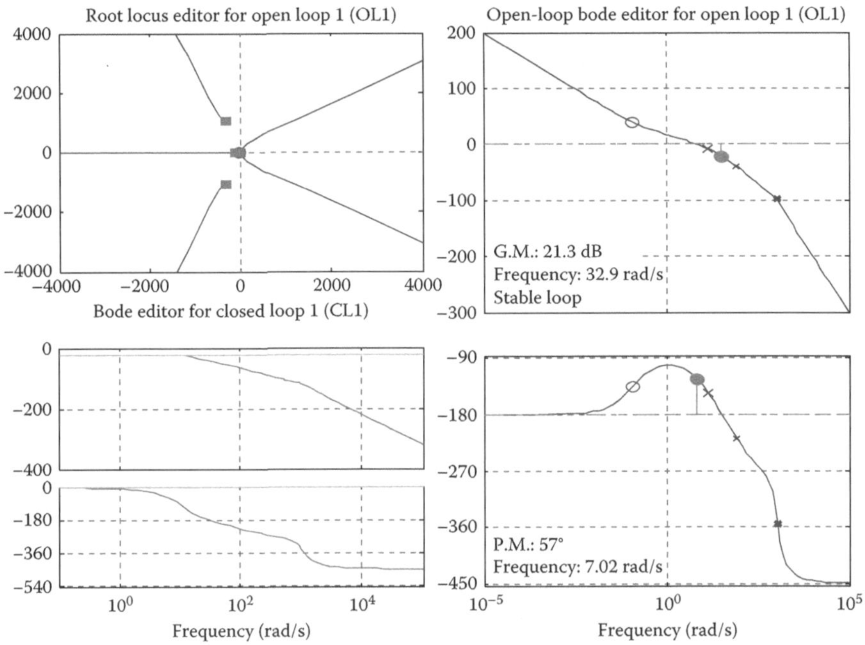

FIGURE 3.12
Frequency domain characteristic of the open-loop control system for the shoulder actuator for Equation 3.56.

The frequency domain characteristic of the open-loop control system for the upper arm actuator is as shown in Figure 3.12. In the right Bode chart, the amplitude margin G.M. is 21.3 dB and the phase margin P.M. is 57°, and the closed loop is a stable loop. Generally, if G.M. is larger than +6 dB and P.M. is larger than 45°, the stability margin is enough. Furthermore, the open-loop root locus shows that all the roots of the closed-loop control system are located in the negative half plane. Thus, the feedback control system by the PI controller $G_c(s)$ is stable. The bandwidth of this closed-loop control system is 7.02 rad/s, that is, 1.11 Hz. If the sinusoidal input demand is less than 1.11 Hz, then the dynamic tracking performance is satisfactory. Otherwise, the tracking performance will be declined. The

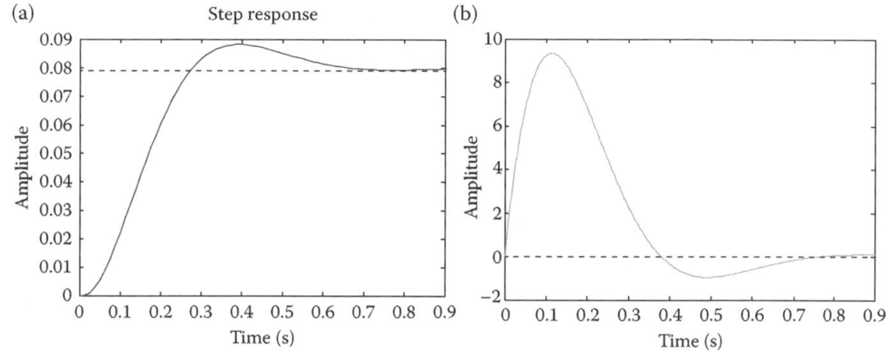

FIGURE 3.13
Performance results of the PI controller design for the shoulder actuator for Equation 3.56. (a) The step response of the upper arm cylinder for the maximal stroke y. (b) The voltage control with respect to step response u.

performance results of the PI controller design for the shoulder actuator is as shown in Figure 3.13. As a step demand is selected as the input, the actual position response of the upper arm cylinder is as shown in Figure 3.13a. The step response for the maximal stroke is regulated to the maximum value 79 mm by the PID controller after 0.8 s. The overshoot is less than 13% and the rising time is less than 0.3 s, which indicates the superior performance of the control design. The voltage control u with respect to step response is shown as Figure 3.13b. The dynamic magnitude of u is less than the control saturation $u_{\max} = 10\,\text{V}$, which satisfies the engineering requirement.

2. If the hydraulic cylinder is retracted, that is, $\dot{y} < 0$, the control parameters and the lag parameter are $k_p = 102$, $k_i = 12$, $b_1 = 0.073$.

   The frequency domain characteristic of the open-loop control system for the upper arm actuator is as shown in Figure 3.14. The amplitude margin G.M. is 18.6 dB, the phase margin P.M. is 49.5°, and the bandwidth of this closed-loop control system is 9.08 rad/s, that is, 1.43 Hz, which is larger than case 1. The performance results of case 2 is shown as Figure 3.15. The steady-state time is 1 s, which has a little weaker performance than case 1. However, the control magnitude is less than 5 V, which denotes that the control consumption of case 2 is more than case 1.

To address different external loads, the required supply pressure (3.55) is different. Thus, from Equations 3.21 and 3.22, the model parameters K_{qa}, K_{qb}, K_{cea}, and K_{ceb} are distinguished. To guarantee enough stability margin of $G_{ol}(s)$ and the similar dynamic performance, the control parameters

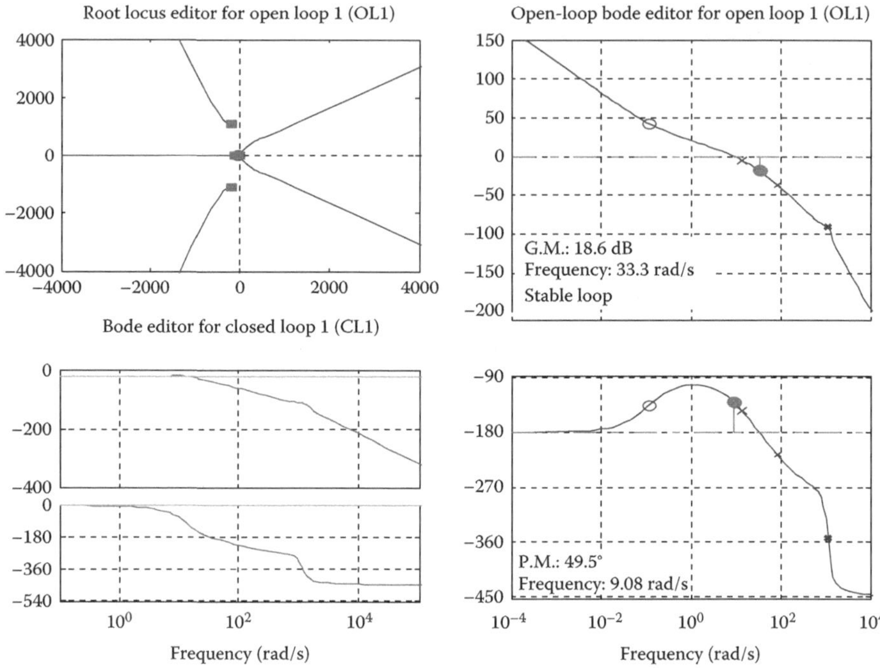

FIGURE 3.14
Frequency domain characteristic of the open-loop control system for the shoulder actuator for Equation 3.57.

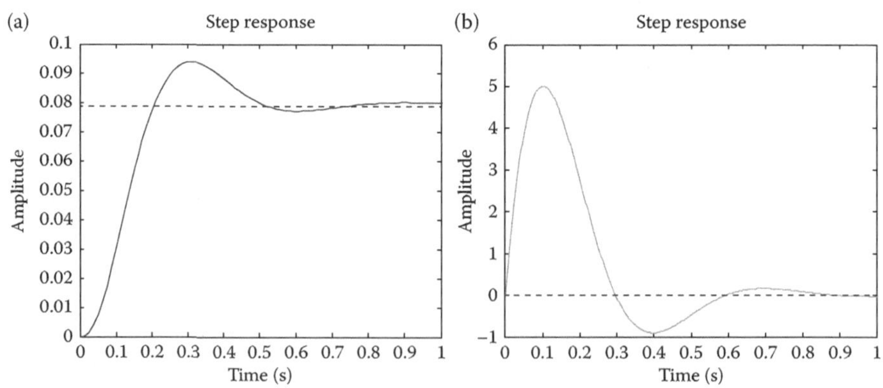

FIGURE 3.15
Performance results of the PI controller design for the shoulder actuator for Equation 3.57. (a) The step response of the upper arm cylinder for the maximal stroke y. (b) The voltage control with respect to step response u.

TABLE 3.7

Control Parameters and Performance Design with Respect to Different Pressures of the Upper Arm Cylinder with Different Loads in the Condition of Extended State

Load Pressure (bar)	K_p/K_i	Bandwidth (rad/s)	Steady Time (s)	Maximum Control (V)
20	150/10	8.92	1	7.42
40	150/10	8.37	1	7.53
60	150/10	7.73	1	7.68
80	170/10	7.74	1	8.70
100	179/21	7.02	0.9	9.34

TABLE 3.8

Control Parameters and Performance Design with Respect to Different Pressures of the Upper Arm Cylinder with Different Loads in the Condition of Retracted State

Load Pressure (bar)	K_p/K_i	Bandwidth (rad/s)	Steady Time (s)	Maximum Control (V)
20	128/15	8.7	1	6.35
40	128/15	9.28	1	6.25
60	119/14	9.31	1	5.82
80	110/13	9.24	1	5.40
100	102/12	9.08	1	5.0

need to be adaptively regulated with the variable external load. Table 3.7 gives the control parameters and performance design with respect to different pressures of the upper arm cylinder with different loads in the condition of extended state.

As the hydraulic cylinder is controlled to be retracted, the control parameters are also switched into the other designed interpolation table as shown in Table 3.8. Obviously, the control parameters in the retracted condition are less than the corresponding parameters in the extended condition because the external load of the retracted cylinder is negative where the gravity of the load mass need not be compensated.

Similarly, the control parameters of the forearm cylinder is also designed like the above two tables in the condition of extended and retracted states.

3.5 Experiment

To verify the PI controller, the experimental bench of the two-DOF robotic arm driven by EHA is set up as shown in Figure 3.16 [11,12]. The two EHAs include two servo valves (Moog D633-R02K01M0NSM2), two double-acting cylinders (Hoerbiger LB6-1610-0080-4M), an axial piston pump (Takako TFH-315), a servo motor (BALDOR BSM63N-375), and a relief valve. The angle

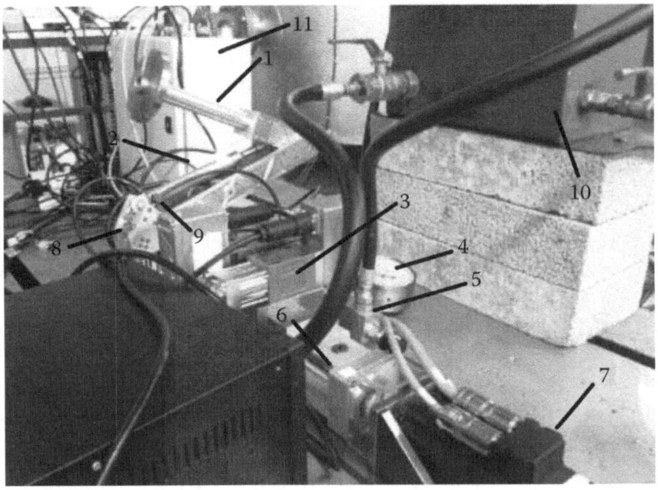

FIGURE 3.16
Experimental equipment (1—robotic arm, 2—hydraulic cylinder, 3—servo valve, 4—pressure gauge, 5—relief valve, 6—fixed displacement pump, 7—servo motor, 8—encoder, 9—pressure sensor, 10—tank, 11—IPC).

TABLE 3.9
Hydraulic Parameters Used in Simulation and Experiment

Parameter	Value	Parameter	Value
C_d	0.62	w	0.024 m
p_s	40 bar	A_a	2.01 cm^2
A_b	1.25 cm^2	ν	0.62
V_t	1.74×10^{-5} m^3	β_e	7×10^8 Pa
K_{sv}	7.9×10^{-5} m/V	$L_{\max}$	58 mm
T_{sv}	12 ms	$u_{\max}$	10 V
K	0	b	2200 N.s/m
C_{tl}	2.5×10^{-11} m^3/(s · Pa)	ρ	850 kg/m^3

of joint is measured by a relative encoder (AVAGO AEDA-3300-BE1), which can obtain the cylinder position by trigonometry computation. The load pressure on each EHA is measured by an optional tension/compression load cell (Burster 8417-6005). The position derivatives is computed by forward Euler method. The control law is executed by industrial personal computer (IPC). The load mass is a disk on the forearm terminal. Some hydraulic parameters of this experimental bench are shown in Table 3.9. The two cylinder position demands are considered as $y_{1d} = 0.5L_{\max}\sin(1.6\pi t)$, $y_{2d} = 0.5L_{\max}\sin(2\pi t)$. According to Tables 3.7 and 3.8, the PI control parameters are designed as $k_p = 150, k_i = 10$, if $\dot{y}_i \geq 0$. Otherwise, $k_p = 128, k_i = 15$.

Figures 3.17 and 3.18 show the position tracking results of the two hydraulic cylinders. The magnitude of two sinusoidal demands is 29 mm, and the two frequencies are 0.8 and 1 Hz. The two dynamic position errors are less than 3 mm, that is, $(y_{id} - y_i)/y_{id} < 10\%$. The position errors of the forearm cylinder is less than that of the upper arm cylinder, since the external load of the upper arm hydraulic actuator is more violent than that of forearm actuator from Figure 3.11. The two control variable u does not exceed the control saturation $\pm 10\,\text{V}$, which means that the designed PI controller can guarantee the stability of the closed-loop feedback control

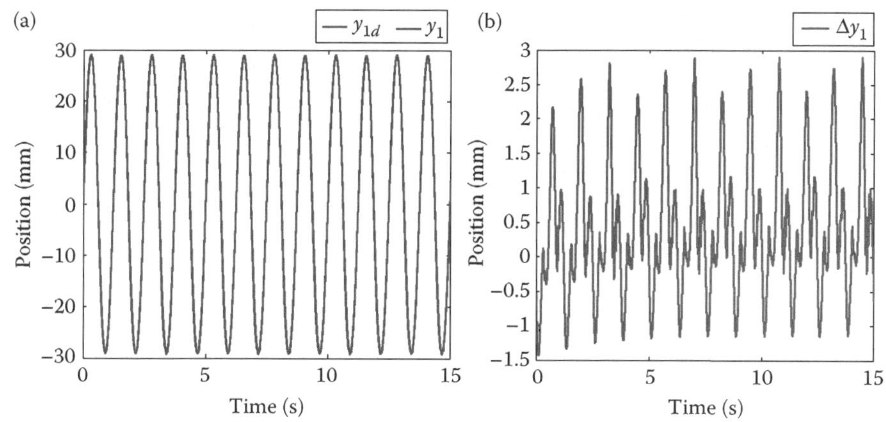

FIGURE 3.17
Position tracking results of the upper arm cylinder. (a) The upper arm position response y_1. (b) The upper arm position error Δy_1.

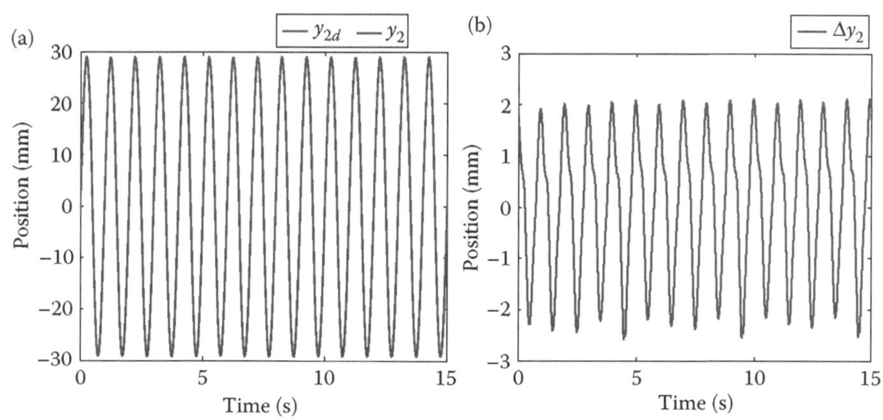

FIGURE 3.18
Position tracking results of the forearm cylinder. (a) The forearm position response y_2. (b) The forearm position error Δy_2.

system of the electro-hydraulic actuator, and the dynamic tracking performance of the cylinder position is satisfactory (Figure 3.19). Figure 3.20 shows the two chamber pressures of the asymmetrical cylinder. These pressures are switched from a small value into the approximate value of the supply pressure, which is less than the supply pressure $p_s = 40\,\text{bar}$.

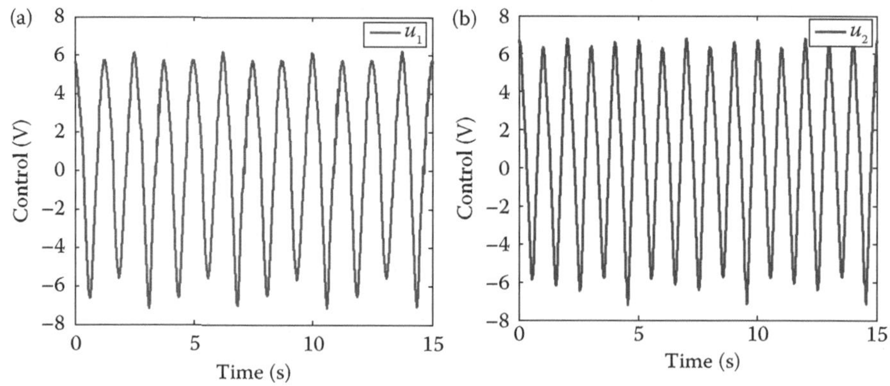

FIGURE 3.19
Dynamic control voltages of two servo valves. (a) The upper arm control variable u_1. (b) The forearm control variable u_2.

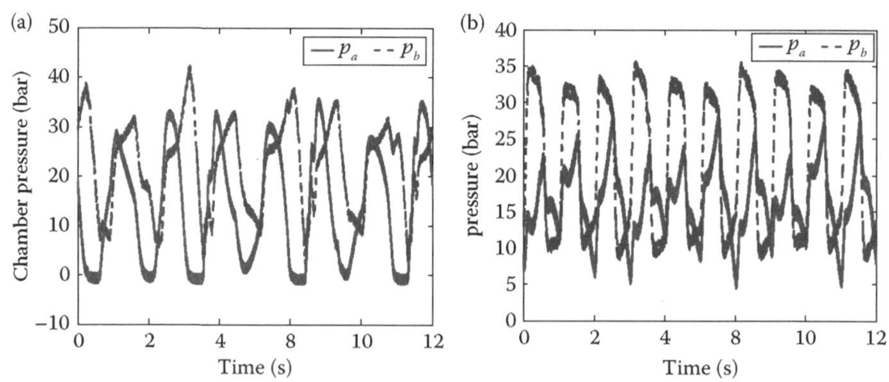

FIGURE 3.20
Two chamber pressures of two cylinders. (a) The two chamber pressures of the upper arm cylinder p_a, p_b. (b) The two chamber pressures of the forearm cylinder p_a, p_b.

4

Robust Control Method

CONTENTS

Some hydraulic parametric uncertainties exist in electro-hydraulic systems. These uncertainties include uncertain linear/nonlinear parameter and uncertain nonlinearity [17], which are caused by unknown viscous damping, load stiffness, variations in control fluid volumes, physical characteristics of valve, bulk modulus, and oil temperature variations [42]. If these uncertainties are not addressed in the control design, the dynamic and steady performance of EHS will be degraded and the robustness of the designed controller will also be reduced. In this chapter, the robust control method of EHS is proposed to handle the parametric uncertainties and compensate their negative effect in real engineering [20]. Thus, the dynamic and steady behaviors are obviously improved, especially the largely unknown load disturbances emerging in the electro-hydraulic actuator.

In this chapter, first, the linear mathematical model of EHS is constructed by flow-pressure model linearization. Second, the external load forces of the two-DOF robot arm are modeled by Lagrange dynamic equation. A bounded linear function is constructed to describe the relationship between load forces and cylinder position. Then the H_∞ suboptimal controller is designed in the output feedback form with structural and parametric uncertainty. Both the simulation and experiment are performed to validate the effectiveness of the developed algorithm.

4.1 Linearized Hydraulic Model Construction

In this section, for convenient discussion, the dynamic cylinder model of EHA is adopted by the linearized hydraulic model mentioned in Equations 2.4 and 2.25. Thus, the linear model of the hydraulic cylinder is shown as follows:

$$y(s) = \frac{\frac{K_{qi}}{A_a}x_v(s) - \frac{1}{A_a^2}\left(K_{cei} + \frac{V_t}{2(1+\upsilon^2)\beta_e}s\right)F_L(s)}{\frac{V_t m_t}{2(1+\upsilon^2)\beta_e A_a^2}s^3 + \left(\frac{m_t K_{cei}}{A_a^2} + \frac{bV_t}{2(1+\upsilon^2)\beta_e A_a^2}\right)s^2 + \left(1 + \frac{bK_{cei}}{A_a^2} + \frac{V_t K}{2(1+\upsilon^2)\beta_e A_a^2}\right)s + \frac{KK_{cei}}{A_a^2}}, \tag{4.1}$$

where y is the displacement of the piston, x_v is the spool position of the servo valve, K_{qi}, $K_{cei}(i = a, b)$ are the flow gain of the servo valve and flow-pressure coefficient, A_a and A_b are the annulus areas of the two chambers ($\upsilon = A_a/A_b$), V_t is the cylinder volume, β_e is the effective bulk modulus, m_t is the load mass, K is the load spring constant, b is the viscous damping coefficient of oil, and F_L is the external load on the hydraulic actuator from the mechanical structure of the two-DOF robotic arm.

Remark 4.1

From Equation 4.1, it is clear that the hydraulic cylinder model is simplified as a third-order linear model. If there exist bounded uncertain parameters such as V_t, K, b, K_q, and K_{ce}, this linear model is appropriately handled by the robust control method, which obtains an H_∞ suboptimal solution to preserve a satisfactory dynamic control performance.

The dynamic characteristic of the servo valve can be described as a second-order linear model mentioned in Equation 2.9 as follows:

$$\ddot{x}_v + 2\zeta_{sv}\omega_{sv}\dot{x}_v + \omega_{sv}^2 x_v = K_{sv}\omega_{sv}^2 u, \tag{4.2}$$

where ζ_{sv} is the damping ratio, ω_{sv} is the natural frequency, K_{sv} is the servo valve gain, and u is the control voltage.

According to Equations 4.1 and 4.2, the whole state-space model of the electro-hydraulic actuator is five. The control input variable is u and the output cylinder position is y.

4.2 Analysis of Parametric Uncertainty

In the linear model of the cylinder (4.1), V_t, b, and K_q are three parametric uncertainties. They are caused by operating conditions, environmental variability, hydraulic characteristic variations, and the linearized model error. The flow gain of the servo valve K_q changes with time due to variable load pressure P_L and component degradation variations. Because of the asymmetry cylinder, the value of K_q is different in cylinder extension or retraction, respectively.

Some hydraulic parameters used in simulation and experiments are shown in Table 4.1. According to Equations 3.16 and 3.21, the load flow gain of valve K_q for the single-rod electro-hydraulic actuator is rewritten as follows:

$$\begin{aligned} K_q &= K_{q0}\sqrt{\frac{p_s - p_l - \Delta p_{vloss} - \Delta p_{tube}}{p_s}}, \\ K_{qa} &= K_q\sqrt{\frac{2}{1+\upsilon^3}}, \\ K_{qb} &= K_q\sqrt{\frac{2(\upsilon(p_s - \Delta p_{vloss} - \Delta p_{tube}) + p_l)}{(p_s - \Delta p_{vloss} - \Delta p_{tube} - p_l)(1+\upsilon^3)}}. \end{aligned} \tag{4.3}$$

The load mass is 1 kg and the motion frequency of the shoulder arm is assumed to be no more than 0.5 Hz. Thus, the load pressure p_L range of the shoulder cylinder can be estimated by Lagrange equation. The maximum value of p_L is 40 bar. If some known parameters are substituted into Equation 4.3, the flow gain ranges of the rod chamber and non-rod chamber are

TABLE 4.1

Hydraulic Parameters Used in Simulation and Experiments

Parameter	Value	Parameter	Value
$x_{v\max}$	0.79 mm	α	0.62
P_s	40 bar	P_r	2 bar
A_a	2.01 cm^2	K_{q0}	0.015 m^2/s
K_{sv}	3.95×10^{-4} m/V	β_e	2×10^8 Pa
ω_{sv}	353.6 rad/s	ζ_{sv}	0.707
Δp_{vloss}	1.4 bar	Δp_{tube}	7.5 bar
L_c	157 mm	$d_{r\max}$	79 mm
m_1	1.772 kg	m_2	0.739 kg

estimated as follows:

$$\begin{aligned}0.012 &\le K_{qa} \le 0.02,\\ 0.019 &\le K_{qb} \le 0.027.\end{aligned} \tag{4.4}$$

So the range of K_q can be considered as $0.012 < K_q < 0.027$. The nominal values $\bar{K}_q$ of K_q is $0.02\,\mathrm{m^2/s}$ and the maximum relative uncertainty p_{K_q} is 0.375 with the relative variation. The flow gain of the servo valve K_q can be represented as follows:

$$K_q = \bar{K}_q(1 + p_{K_q}\delta_{K_q}), \quad -1 \le \delta_{K_q} \le 1. \tag{4.5}$$

Because the dynamic change of volume in two chambers is neglected in Equation 2.1, the cylinder volume V_t is not accurate. The actual volume can be described as follows:

$$\begin{cases} V_a = V_{a0} + A_a y \\ V_b = V_{b0} - A_b y \end{cases}. \tag{4.6}$$

If the following property holds:

$$\begin{aligned}&V_{a0} = V_{b0} = V_t/2,\\ &|A_a y| \ll V_t, \quad |A_b y| \ll V_t,\end{aligned} \tag{4.7}$$

then flow-pressure continuous equation of the cylinder is available. However, the initial volumes of two chambers are unknown, which means perhaps not in the center of the cylinder chamber. So the parameter V_t can be represented as a parametric uncertainty. The initial volumes of two chambers are estimated as follows:

$$\begin{cases} (L_c - d_{r\max})A_a/2 \le V_{a0} \le (L_c + d_{r\max})A_a/2 \\ (L_c - d_{r\max})A_b/2 \le V_{b0} \le (L_c + d_{r\max})A_b/2 \end{cases}. \tag{4.8}$$

After the known parameters are substituted into Equation 4.8, the nominal values $\bar{V}_t$ of V_t is $2.85\text{e-}5\,\mathrm{m^3}$ and the maximum relative uncertainty p_{V_t} is 0.663 with the relative variation δ_{V_t}. The cylinder volume V_t can be represented as follows:

$$V_t = \bar{V}_t(1 + p_{V_t}\delta_{V_t}), \quad -1 \le \delta_{V_t} \le 1. \tag{4.9}$$

The viscous damping coefficient of the cylinder b can be estimated by the experiment test. The value is 5000 or 2000 Ns/m in cylinder extension or retraction, respectively. So the nominal values $\bar{b}$ of b is 3500 Ns/m and the maximum relative uncertainty p_b is 0.428 with the relative variation δ_b. The viscous damping coefficient b can be represented as follows:

$$b = \bar{b}(1 + p_b\delta_b), \quad -1 \leq \delta_b \leq 1. \tag{4.10}$$

Now the external load F_L is the maximum disturbance in the dynamic characteristic of the cylinder. It can be considered as an unmodeled disturbance. In this chapter, the quantitative relationship between the external load F_L and the position of the cylinder is analyzed to describe the structural uncertainty. According to Equation 3.43, two torques on the upper arm and forearm T_u and T_f are obtained by Lagrange method. Then, two external load forces F_{Lu}, F_{Lf} that change with the variation of two joint angles are computed by

$$\begin{cases} F_{Lu}(\theta_1, \theta_2) = \dfrac{T_u(\theta_1, \theta_2)}{l_1(\theta_1)} \\ F_{Lf}(\theta_1, \theta_2) = \dfrac{T_f(\theta_1, \theta_2)}{l_2(\theta_2)} \end{cases}, \tag{4.11}$$

where $l_i(\theta_i)(i = 1, 2)$ are the force arms of the external load forces.

To consider the effect of the external load in the linear model (4.1) of EHS, two fictitious proportional gains are defined as follows:

$$\begin{cases} K_{F_{Lu}}(\theta_1) = \dfrac{F_{Lu}(\theta_1, \theta_2)}{c_1(\theta_1)} = \dfrac{T_u(\theta_1, \theta_2)}{l_1(\theta_1)c_1(\theta_1)} \\ K_{F_{Lu}}(\theta_2) = \dfrac{F_{Lu}(\theta_1, \theta_2)}{c_2(\theta_2)} = \dfrac{T_u(\theta_1, \theta_2)}{l_1(\theta_1)c_2(\theta_2)} \\ K_{F_{Lf}}(\theta_1) = \dfrac{F_{Lf}(\theta_1, \theta_2)}{c_1(\theta_1)} = \dfrac{T_f(\theta_1, \theta_2)}{l_2(\theta_2)c_1(\theta_1)} \\ K_{F_{Lf}}(\theta_2) = \dfrac{F_{Lf}(\theta_1, \theta_2)}{c_2(\theta_2)} = \dfrac{T_f(\theta_1, \theta_2)}{l_2(\theta_2)c_2(\theta_2)} \end{cases}, \tag{4.12}$$

where $c_i(\theta_i)(i = 1, 2)$ are the dynamic lengths of the cylinder position.

The dynamic ranges of $K_{F_{Lu}}$ and $K_{F_{Lf}}$ in a duration are shown in Figure 4.1. Here, two fictitious proportional gains $K_{F_{Lu}}(\theta_1)$ and $K_{F_{Lf}}(\theta_2)$ are larger than the other two gains, respectively. It means that the external load F_{Lu} is influenced by the shoulder angle θ_1 more than θ_2. It is similar to F_{Lf}.

So the structural uncertainty caused by F_L can be described as follows:

$$\begin{cases} F_{Lu} = K_{F_{Lu}}(\theta_1)y_1 \\ F_{Lf} = K_{F_{Lf}}(\theta_2)y_2 \end{cases}. \tag{4.13}$$

The nominal values $\bar{K}_{F_{Lu}}$, $\bar{K}_{F_{Lf}}$ of $K_{F_{Lu}}$ and $K_{F_{Lf}}$ are $0\,\text{N/m}$, which means the no-load motion. The maximum relative uncertainties $p_{K_{FLu}}$ and $p_{K_{FLf}}$ are

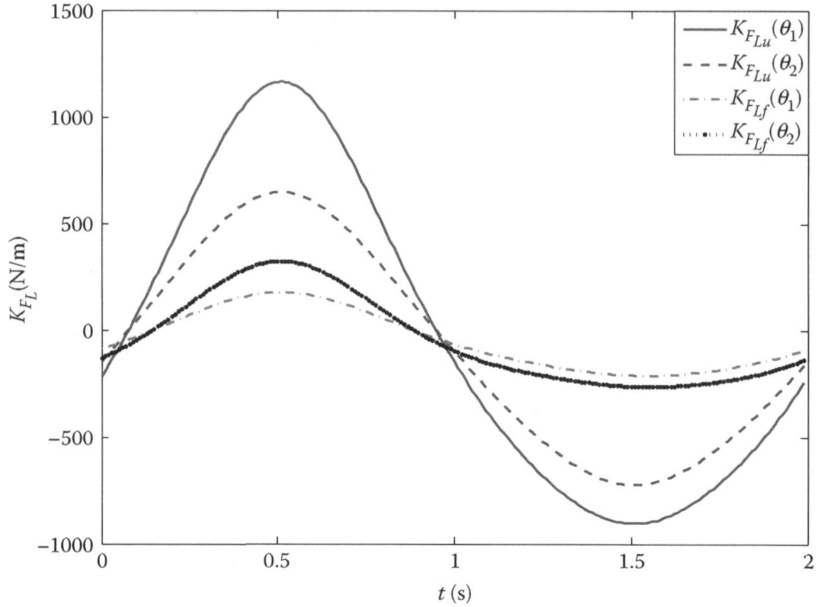

FIGURE 4.1
Dynamic ranges of $K_{F_{Lu}}$ and $K_{F_{Lf}}$ in one motion duration.

1200 and 300 with the relative variations $\delta_{K_{FLu}}$ and $\delta_{K_{FLf}}$. The two fictitious proportional gains and can be represented as follows:

$$\begin{cases} K_{F_{Lu}} = \bar{K}_{F_{Lu}}(1 + p_{K_{FLu}}\delta_{K_{FLu}}), & -1 \leq \delta_{K_{FLu}} \leq 1 \\ K_{F_{Lf}} = \bar{K}_{F_{Lf}}(1 + p_{K_{FLf}}\delta_{K_{FLf}}), & -1 \leq \delta_{K_{FLf}} \leq 1 \end{cases}. \tag{4.14}$$

Remark 4.2

From Equations 4.5, 4.9, and 4.10, the three parametric uncertainties K_q, V_t, and b are adopted by the multiplication uncertain expressions. However, K_{F_L} is adopted by the addition uncertain expression.

4.3 Robust Model Construction

The parameters K_q, V_t, and b may be represented as an upper linear fractional transformation (LFT) [14,39], that is, $K_q = F_u(M_{K_q}, \delta_{K_q})$, $V_t = F_u(M_{V_t}, \delta_{V_t})$, $b = F_u(M_b, \delta_b)$, where

$$M_{K_q} = \begin{bmatrix} 0 & \bar{K}_q \\ p_{K_q} & \bar{K}_q \end{bmatrix},$$
$$M_{V_t} = \begin{bmatrix} -p_{V_t} & 1/\bar{V}_t \\ -p_{V_t} & 1/\bar{V}_t \end{bmatrix}, \tag{4.15}$$
$$M_{B_e} = \begin{bmatrix} 0 & \bar{B}_e \\ p_{B_e} & \bar{B}_e \end{bmatrix}.$$

The structural uncertainty parameters $K_{F_{Lu}}$, $K_{F_{Lf}}$ can be represented as $K_{F_{Lu}} = F_u(M_{K_{F_{Lu}}}, \delta_{K_{F_{Lu}}})$, $K_{F_{Lf}} = F_f(M_{K_{F_{Lf}}}, \delta_{K_{F_{Lf}}})$, where

$$M_{K_{F_{Lu}}} = \begin{bmatrix} 0 & 1 \\ p_{K_{F_{Lu}}} & \bar{K}_{F_{Lu}} \end{bmatrix},$$
$$M_{K_{F_{Lf}}} = \begin{bmatrix} 0 & 1 \\ p_{K_{F_{Lf}}} & \bar{K}_{F_{Lf}} \end{bmatrix}. \tag{4.16}$$

Taking the shoulder hydraulic actuator for example, the robust model with parametric and structural uncertainties is shown in Figure 4.2. If some vectors are defined as the state variable vector $X = [x_1, x_2, x_3, x_4, x_5]^T$, the control inputs $u(t)$ is the voltage of the servo valve, the vector of exogenous inputs $w(t) = [w_1,\ w_2,\ w_3,\ w_4,\ w_5,\ w_6]^T$, the vector of measurements $y(t)$, the regulated output vector $z(t) = [z_1,\ z_2,\ z_3,\ z_4,\ z_5,\ z_6]^T$, then the linear uncertain state-space model is described as follows:

$$\begin{cases} \dot{X}(t) = AX(t) + B_1W(t) + B_2u(t) \\ Z(t) = C_1X(t) + D_{11}W(t) \\ Y(t) = C_2X(t) + Y_d(t) \end{cases}, \tag{4.17}$$

where

$$A = \begin{bmatrix} 0 & 1 & 0 & 0 & 0 \\ -\omega_{sv}^2 & -2\zeta_{sv}\omega_{sv} & 0 & 0 & 0 \\ 0 & 0 & 0 & 1 & 0 \\ 0 & 0 & 0 & 0 & 1 \\ \frac{4\bar{K}_q\beta_eA_a}{\bar{V}_tm_{1f}} & 0 & -\frac{4\beta_e\bar{K}_{F_{Lu}}K_{ce}}{\bar{V}_tm_{1f}} & -\frac{4\beta_eA_a^2}{\bar{V}_tm_{1f}} - \frac{\bar{b}\beta_eK_{ce}}{\bar{V}_tm_{1f}} - \frac{\bar{K}_{F_{Lu}}}{m_{1f}} & \frac{-4K_{ce}\beta_e}{\bar{V}_t} - \frac{\bar{b}}{m_{1f}} \end{bmatrix}, \tag{4.18}$$

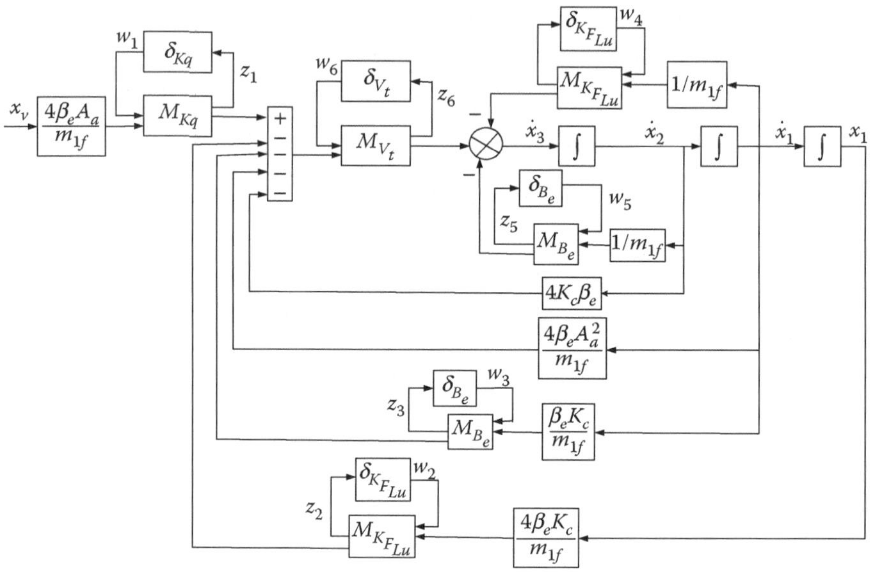

FIGURE 4.2
Robust model with parametric and structural uncertainties.

$$B_1 = \begin{bmatrix} 0 & 0 & 0 & 0 & 0 & 0 \\ 0 & 0 & 0 & 0 & 0 & 0 \\ 0 & 0 & 0 & 0 & 0 & 0 \\ 0 & 0 & 0 & 0 & 0 & 0 \\ \frac{p_{K_q}}{\bar{V}_t} & -\frac{p_{K_{F_{Lu}}}}{\bar{V}_t} & -\frac{p_{B_e}}{\bar{V}_t} & -p_{K_{F_{Lu}}} & -p_{B_e} & -p_{v_t} \end{bmatrix}, \tag{4.19}$$

$$B_2 = \begin{bmatrix} 0 & K_{sv}\omega_{sv}^2 & 0 & 0 & 0 \end{bmatrix}^T, \tag{4.20}$$

$$C_1 = \begin{bmatrix} \frac{4\bar{K}_q\beta_e A_a}{m_{1f}} & 0 & 0 & 0 & 0 \\ 0 & 0 & \frac{4\beta_e K_c}{m_{1f}} & 0 & 0 \\ 0 & 0 & 0 & \frac{\bar{b}\beta_e K_c}{m_{1f}} & 0 \\ 0 & 0 & 0 & \frac{1}{m_{1f}} & 0 \\ 0 & 0 & 0 & 0 & \frac{\bar{B}_e}{m_{1f}} \\ \frac{4\bar{K}_q\beta_e A}{\bar{V}_t m_{1f}} & 0 & -\frac{4\beta_e\bar{K}_{F_{Lu}}K_c}{\bar{V}_t m_{1f}} & -\frac{4\beta_e A_a^2}{\bar{V}_t m_{1f}} - \frac{\bar{B}_e\beta_e K_c}{\bar{V}_t m_{1f}} & \frac{-4K_c\beta_e}{\bar{V}_t} \end{bmatrix}, \tag{4.21}$$

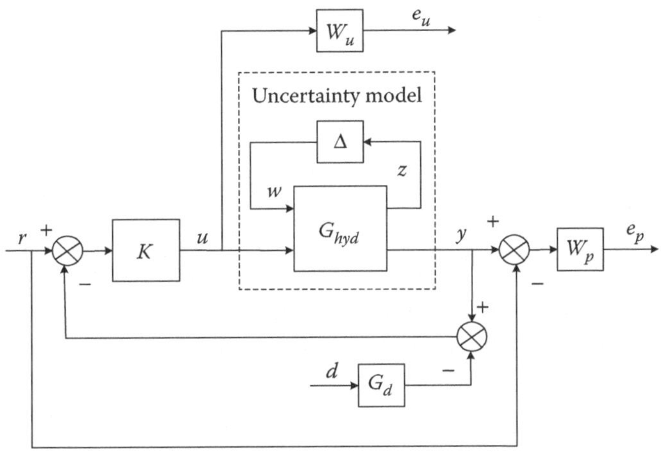

FIGURE 4.3
Block diagram of the closed-loop system with robust performance requirements.

$$D_{11} = \begin{bmatrix} 0 & 0 & 0 & 0 & 0 & 0 \\ 0 & 0 & 0 & 0 & 0 & 0 \\ 0 & 0 & 0 & 0 & 0 & 0 \\ 0 & 0 & 0 & 0 & 0 & 0 \\ 0 & 0 & 0 & 0 & 0 & 0 \\ \frac{p_{K_q}}{\bar{V}_t} & -\frac{p_{K_{FLu}}}{\bar{V}_t} & -\frac{p_{B_e}}{\bar{V}_t} & 0 & 0 & -p_{v_t} \end{bmatrix}, \tag{4.22}$$

$$C_2 = \begin{bmatrix} 0 & 0 & 1 & 0 & 0 \end{bmatrix}. \tag{4.23}$$

Here, the influence of load stiffness is neglected, that is, $K = 0$. The measurement noise $Y_d(t)$ can be described in frequency domain as follows:

$$Y_d(s) = G_d(s)N(s), \quad G_d(s) = 0.006 \times \frac{1 + 0.12s}{1 + 0.001s}, \tag{4.24}$$

where $N(s)$ is the noise input.

The block diagram of the closed-loop system with robust performance requirements is shown in Figure 4.3. In this figure, the linear uncertain state-space model is described as block G_{hyd}, the block $\Delta = diag(\delta_{K_q}, \delta_{V_t}, \delta_{B_e}, \delta_{K_{FLu}})$, r is the demand displacement of the cylinder input, d is the measurement noise input, and K is the robust controller. The weighting functions W_p and W_u are used to reflect the relative significance of the performance requirement.

Remark 4.3

The block diagram of the closed-loop system (Figure 4.3) involves two inputs r and d, and two outputs e_p and e_u, where e_p is the tracking error of the cylinder position and e_u is the control variable. The objective of the robust controller is to design K to guarantee the tracking error $e_p = W_p(r - y)$ achieving the required performance under the control saturation constraint of the control variable e_u and the existed external disturbance d.

4.4 Robust Controller Design

The goal of the controller design is to satisfy the robust stability and performance in the case of the parametric and structural uncertainties. In addition, the controller should reduce the output sensitivity due to the presence of measurement noise.

4.4.1 Analysis of Open-Loop System

If the structural uncertainty model F_L (4.13) is embedded in the linear EHS model (4.1) and (4.2), then the open-loop system of the EHS model from the control variable u to the output cylinder position y is given by

$$\frac{Y(s)}{u(s)} = \frac{\frac{K_{qi}}{A_a}\frac{K_{sv}\omega_{sv}^2}{s^2+2\zeta_{sv}\omega_{sv}s+\omega_{sv}^2}}{\begin{array}{l}\left(\frac{V_t m_t}{2(1+\upsilon^2)\beta_e A_a^2}s^3 + \left(\frac{m_t K_{cei}}{A_a^2} + \frac{bV_t}{2(1+\upsilon^2)\beta_e A_a^2}\right)s^2 + \right. \\ \left.\left(1 + \frac{bK_{cei}}{A_a^2} + \frac{V_t K}{2(1+\upsilon^2)\beta_e A_a^2} + \frac{V_t K_{F_{Li}}}{2A_a^2(1+\upsilon^2)\beta_e}\right)s + \frac{KK_{cei}}{A_a^2}\right) / \left(1 + \frac{K_{cei}K_{F_{Li}}}{A_a^2}\right)\end{array}}, \tag{4.25}$$

where $K_{F_{Li}}(i = u, f)$ are two fictitious proportional gains defined in Equation 4.12.

By referring to the open-loop control system analysis in Section 3.1, G_{open} needs to be analyzed since the three performance indexes, that is, the magnitude margin A_m, the phase margin P_m, and the system bandwidth ω_c, are reflected in the open-loop system. The frequency response of the open-loop system with uncertainties is shown in Figure 4.4. The magnitude is little different between perturbed parameter and nominal parameter. But the phase is significantly different, which is mainly caused by the structural uncertainty parameter $K_{F_{Lu}}$. Owing to the presence of $K_{F_{Lu}}$, the stable phase of the open-loop system will lead or lag approximately 40° compared to the nominal model in the middle frequency stage. So the important objective of robust controller design is to compensate the phase margin of the system.

It should be noted that, in a general condition, the external load $F_L(t)$ is a typical disturbance input of the EHS. The linear EHS model (4.1) and (4.2) is difficult to simplify into Equation 4.25 due to the two system inputs u and F_L. Thus, according to the analysis of the dynamic ranges of $K_{F_{Lu}}$ and $K_{F_{Lf}}$ in a

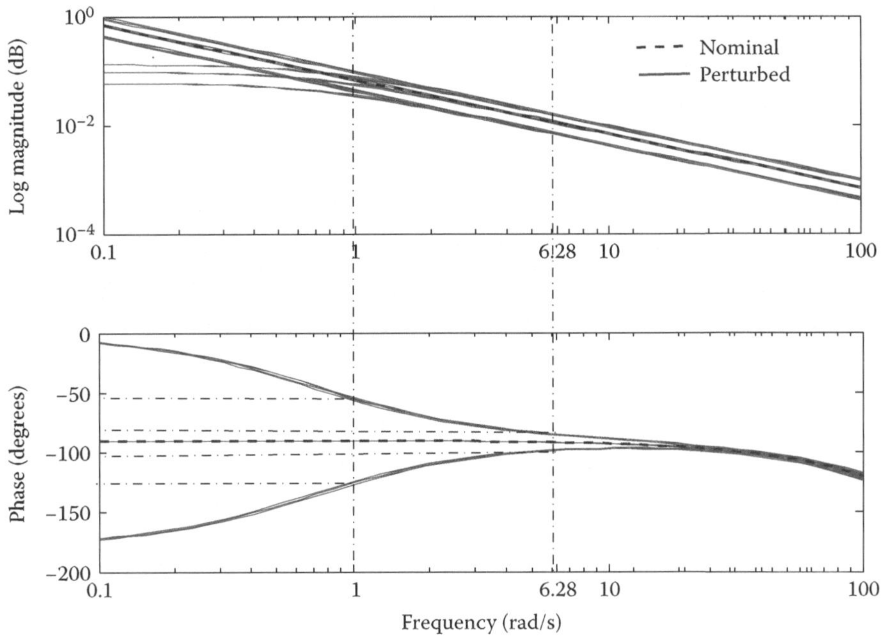

FIGURE 4.4
Frequency response of the open-loop system with varying uncertainty parameters K_q, V_t, b, and $K_{F_{Lu}}$.

duration in Section 4.2, the simplified linearized model (4.13) can be substituted into Equation 4.1, and the open-loop system of the EHS model (4.25) is obtained.

4.4.2 Weight Function Design

The weighting functions W_p and W_u are iteratively designed according to robust performance requirement and control constraint. If W_p is larger, the dynamic tracking performance of EHS is required to be higher. But the robust controller is more difficult to find a feasible solution in this case. Otherwise, the robust controller is easily designed. If W_u is larger, the control constraint is rigorous, which means the control variable u is bounded in sufficiently small value. In this case, the robust controller is also difficult to design. The general form of W_p is designed as follows [15]:

$$W_p = \frac{s/M_s + \omega_c}{s + \omega_c A_e}. \tag{4.26}$$

Here, A_e is the maximum steady accuracy requirement, which is selected as 0.01. It means the relative steady error between system output and demand input is less than 1% in the lower-frequency domain. M_s is the peak of sensitivity, which is selected as 2. ω_c is the prospective bandwidth of the closed loop, which is selected as 2π according to the bandwidth range of the demand position input. The weighting function W_u is selected as 6×10^{-4}, which can satisfy the maximum control voltage $\pm 10\,\text{V}$ of the servo valve in the following iteration.

Theorem 4.1 ([67] Small Gain Theorem)

Consider the general feedback system in Figure 4.5, which includes two cross-linked subsystems H_1 and H_2. If there exist four constants γ_1, γ_2, c_1, and c_2 such that

$$\begin{cases} \|H_1 e_1\| \le \gamma_1 \|e_1\| + c_1 \\ \|H_2 e_2\| \le \gamma_2 \|e_2\| + c_2 \end{cases}, \tag{4.27}$$

where $\forall e_1, e_2 \in L_p$, and $\gamma_1 \gamma_2 < 1$, then there exist two sufficiently large constants γ and c such that

$$\left\| \begin{bmatrix} e_1 \\ e_2 \end{bmatrix} \right\| \le \gamma \left\| \begin{bmatrix} u_1 \\ u_2 \end{bmatrix} \right\| + c, \tag{4.28}$$

where two inputs $u_1, u_2 \in L_p$.

Proof. According to Figure 4.5, we can see that

$$\begin{aligned} y_1 &= H_1 e_1, \quad e_1 = u_1 - y_2, \\ y_2 &= H_2 e_2, \quad e_2 = u_2 + y_1. \end{aligned} \tag{4.29}$$

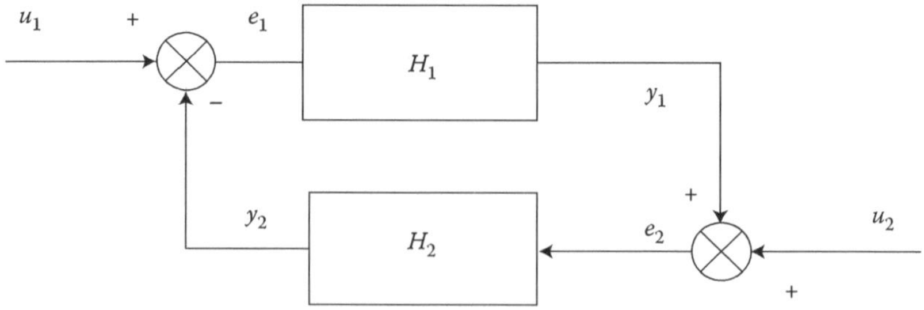

FIGURE 4.5
Cross-linked feedback system.

Thus, if two errors e_1 and e_2 are defined as the system outputs, and u_1, u_2 are system inputs, then the following condition is satisfied:

$$\begin{cases} e_1 = u_1 - H_2 e_2 \\ e_2 = u_2 + H_1 e_1 \end{cases}. \tag{4.30}$$

So, from triangular inequality rule, the following property holds:

$$\begin{cases} \|H_1 e_1\| = \|u_2 - e_2\| \geq \|e_2\| - \|u_2\| \\ \|H_2 e_2\| = \|u_1 - e_1\| \geq \|e_1\| - \|u_1\| \end{cases}. \tag{4.31}$$

Substituting Equation 4.31 into Equation 4.27, we can see that

$$\begin{cases} \|e_2\| \leq \gamma_1 \|e_1\| + \|u_2\| + c_1 \\ \|e_1\| \leq \gamma_2 \|e_2\| + \|u_1\| + c_2 \end{cases}. \tag{4.32}$$

To address Equation 4.32, if the two sides of two equalities are multiplied by γ_1, γ_2, respectively, and are incorporated together, we can obtain

$$\begin{cases} \|e_1\| \leq \dfrac{1}{1-\gamma_1\gamma_2}(\|u_1\| + \gamma_2 \|u_2\| + c_2 + \gamma_2 c_1) \\ \|e_2\| \leq \dfrac{1}{1-\gamma_1\gamma_2}(\|u_2\| + \gamma_1 \|u_1\| + c_1 + \gamma_1 c_2) \end{cases}. \tag{4.33}$$

If two constants γ and c are defined as follows:

$$\gamma = \left\| \begin{bmatrix} \dfrac{1}{1-\gamma_1\gamma_2} & \dfrac{\gamma_2}{1-\gamma_1\gamma_2} \\ \dfrac{\gamma_1}{1-\gamma_1\gamma_2} & \dfrac{1}{1-\gamma_1\gamma_2} \end{bmatrix} \right\|, \quad c = \left\| \begin{bmatrix} \dfrac{c_2 + \gamma_2 c_1}{1-\gamma_1\gamma_2} \\ \dfrac{c_1 + \gamma_1 c_c}{1-\gamma_1\gamma_2} \end{bmatrix} \right\|, \tag{4.34}$$

then Equation 4.28 is satisfied. ■

Remark 4.4

Theorem 5.1 denotes that if the feedforward subsystem and feedback subsystem H_1 and H_2 are L stable, and the gain γ_1, γ_2 are bounded by $\gamma_1\gamma_2 < 1$, then the closed-loop system is a bounded-input-bounded-output stable system.

According to Theorem 5.1 and Remark 4.4, if the nominal closed-loop system of EHS G_c^n is satisfied as $\|G_c^n\| < 1$, then the closed-loop system may be

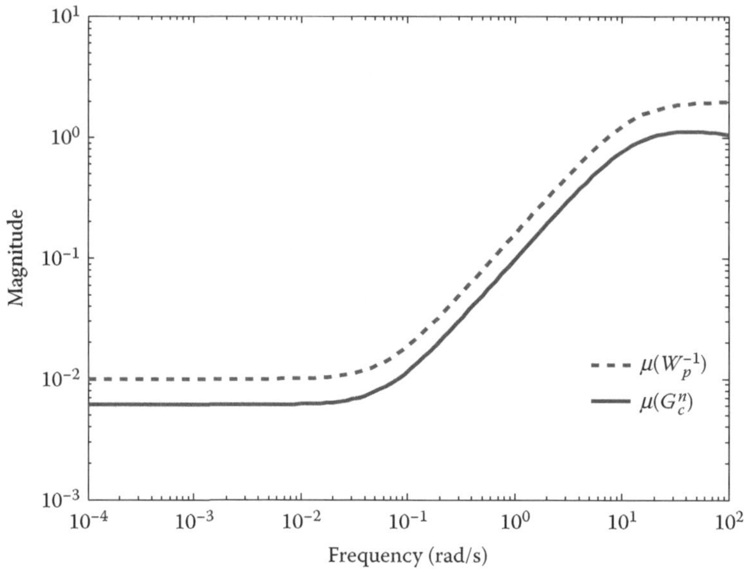

FIGURE 4.6
Singular values of the inverse function $\mu^{-1}(W_p)$ and the nominal closed-loop system $\mu(G_c^n)$.

stable. Furthermore, considering the weight performance function W_p, if the following rule holds, then the closed-loop system is stable:

$$\left\| W_p G_c^n \right\|_{\infty} \leq \mu(W_p)\,\mu(G_c^n) < 1, \tag{4.35}$$

where $\mu(\cdot)$ represents the singular values of the subsystem.

For convenient expression, Equation 4.35 is rewritten as follows:

$$\mu(G_c^n) < \mu^{-1}(W_p). \tag{4.36}$$

Then the singular values of the inverse function $\mu^{-1}(W_p)$ and the nominal closed-loop system $\mu(G_c^n)$ are shown in Figure 4.6. As shown in Figure 4.6, the dynamic variety $\mu(G_c^n)$ is lower than $\mu^{-1}(W_p)$. So Equation 4.35 is satisfied.

4.4.3 Robust Controller Design

The robust controller will solve two problems: robust stability and robust performance requirement [4,10], which can be described in Figure 4.7. The left figure shows the robust stability of the controller K under the parametric uncertainty Δ. In this case, u and w are two inputs and the required performance z, the position response y, and the control performance e_u are three

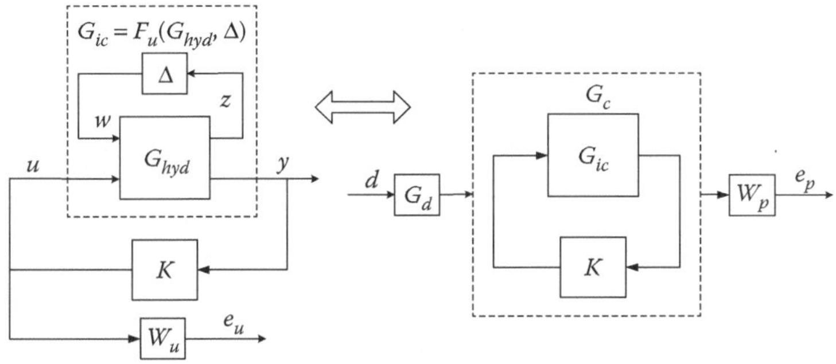

FIGURE 4.7
Block diagram for the description of robust stability and robust performance.

outputs. After the robust stability is discussed, the robust performance of the controller K is analyzed in the right figure. In this case, d is the input and the dynamic tracking performance of position e_p is the output. The robust stability of the designed controller is equivalent to the H_∞ suboptimal control problem as follows.

The subsystem $G_{ic} = F_u\left(G_{hyd}, \Delta\right)$ with the parametric uncertainty is constructed as a cross-link together with the controller K. According to Theorem 5.1, a stabilizing controller K needs to be found such that the H_∞ norm of the closed-loop transfer function is less than a given positive number γ, that is,

$$\|F_l\left(G_{ic}, K\right)\|_\infty = \left\|F_l\left(F_u\left(G_{hyd}, \Delta\right), K\right)\right\|_\infty < \gamma, \tag{4.37}$$

where $\gamma > \gamma_0 := \underset{stabilizing}{\min_K} \|F_l\left(G_{ic}, K\right)\|_\infty$.

The robust performance requirement represents two performances, that is, the tracking error of position response e_p and the control variable e_u. Both of them are constrained by the weight functions W_p and W_u. Thus, if the sensitivity and complementary sensitivity functions of the closed-loop feedback control system satisfies the following constraint:

$$\left\|\begin{bmatrix} W_p(I + G_{ic}K)^{-1}G_d \\ W_uK(I + G_{ic}K)^{-1} \end{bmatrix}\right\|_\infty < 1, \tag{4.38}$$

then the required robust performance is achieved.

The robustness analysis for different uncertainty parameters is given in Figures 4.8 and 4.9. The maximum robust stability bound with varying uncertainty parameters K_q, V_t, b, and $K_{F_{Lu}}$ from the relative uncertainties $\Delta = -I$ to $\Delta = I$. Figure 4.8 shows the upper bound and lower bound of the singular

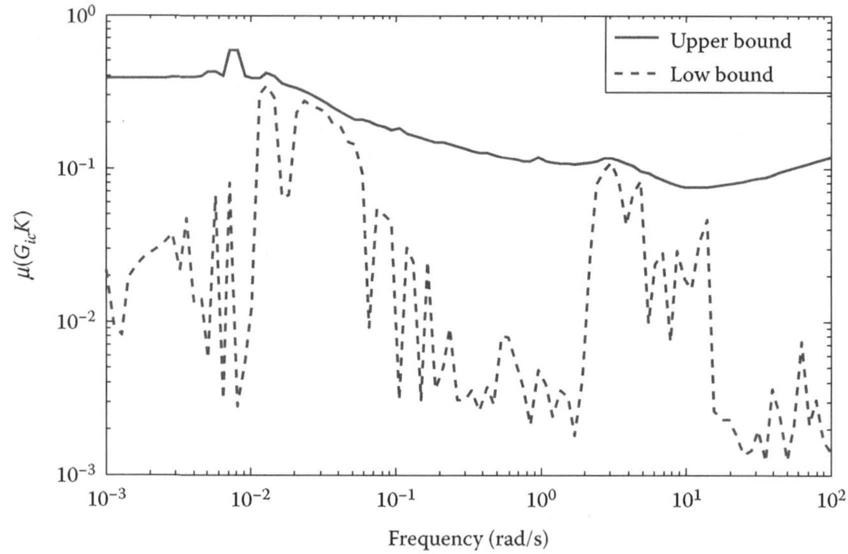

FIGURE 4.8
Maximum robust stability bound with varying uncertainty parameters K_q, V_t, b, and $K_{F_{Lu}}$.

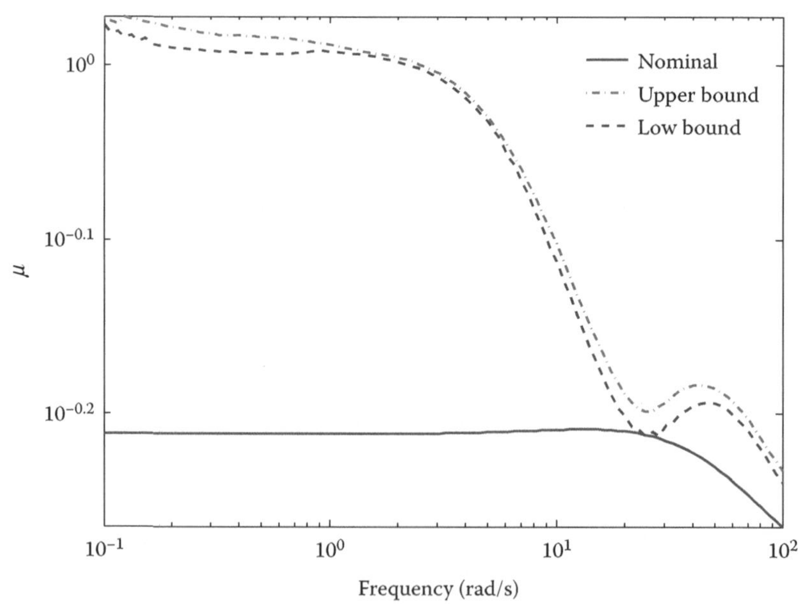

FIGURE 4.9
Maximum robust performance bound with varying uncertainty parameters K_q, V_t, b, and $K_{F_{Lu}}$.

value of $G_{ic}K$, where the parametric uncertainty Δ is involved in G_{ic}. From this figure, $\mu(G_{ic}K) < 1$, which denotes that the feedback control system has the robust stability. No matter how the uncertain parameters K_q, V_t, b, and $K_{F_{Lu}}$ changes, the closed-loop system is stable as shown in Figure 4.7. Different from Figure 4.8, Figure 4.9 gives the singular values between $\mu(G_{sen}G_d)$ and $\mu(1 - G_{sen})$, where G_{sen} is the sensitivity function of the closed-loop system and $1 - G_{sen}$ is the complementary sensitivity function.

In Figure 4.8, the singular value of the closed loop perturbed by uncertainty parameters is 0.37, which is less than 1. It shows that the closed-loop system G_c is stable and its relative uncertainties Δ can be extended to 1/0.37. However, in Figure 4.9, the maximum singular value of the upper bound for $\mu(G_{sen}G_d)$ or $\mu(1 - G_{sen})$ shows that the robust performance requirement is 1.06. It is a little more than 1 in the low frequency stage. It means the steady trace accuracy may not be excellent. But in the middle frequency stage 4–20 rad/s, the maximum singular value is less than 1. It means that the dynamic trace performance and robustness margin meet the requirement [46].

Then the frequency domain result of the designed robust controller is shown in Figure 4.10. After iterative design, the robust controller is seven orders. Taking the realistic project into account, the controller should be reduced into lower order by the balanced truncation method [57]. From Figure 4.10, we can see that reduced order controller is very similar to the

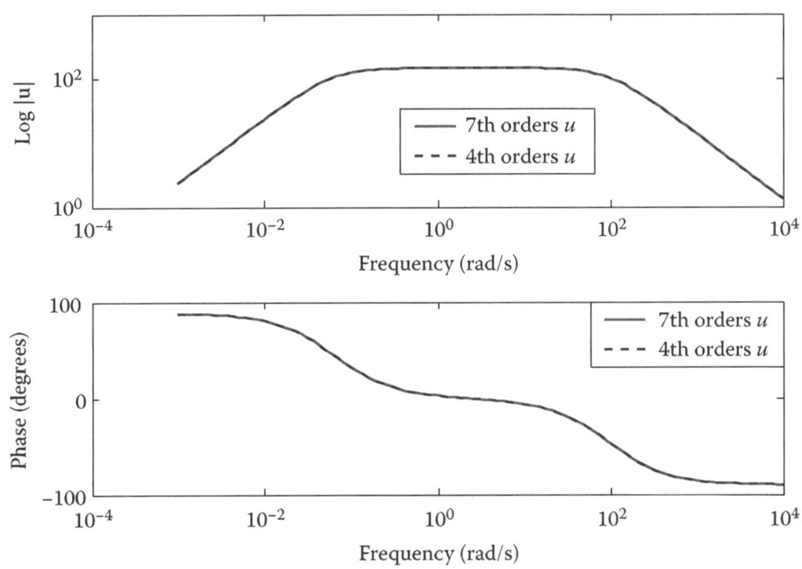

FIGURE 4.10
Frequency domain result of the designed robust controller.

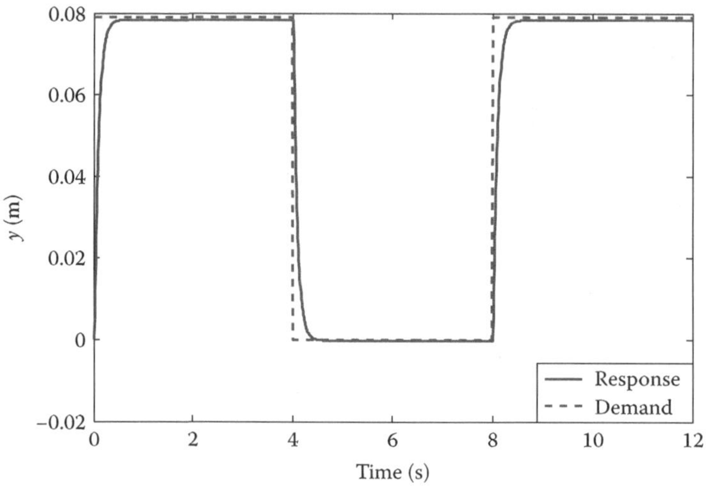

FIGURE 4.11
Cylinder position response in time domain by the designed robust controller.

original seventh-order controller. Thus, the fourth-order control is reasonable. At last, the controller is simplified to four orders as follows:

$$K_i(s) = \frac{n_4 s^4 + n_3 s^3 + n_2 s^2 + n_1 s + n_0}{d_4 s^4 + d_3 s^3 + d_2 s^2 + d_1 s + d_0}\left(y_{id} - y_i\right), \quad i = 1, 2, \tag{4.39}$$

where $n_4 = 0.053$, $n_3 = 1.049$, $n_2 = 155.8$, $n_1 = 5376.2$, $n_0 = 7.6$, $d_4 = 1$, $d_3 = 146.2$, $d_2 = 1938.3$, $d_1 = 6198.7$, and $d_0 = 77.4$, y_{id} is the demand input, and y_i is the measurement output, which can be computed by triangle geometry for the upper arm and forearm hydraulic actuators.

After the robust controller is obtained, the cylinder position response in time domain is given by Figure 4.11. The step demand is a pulse input, where the magnitude is ± 79 mm and the period is 4 s. The response time of the cylinder position is 0.4 s and the steady-state error is less than 1 mm. This denotes that the robust controller can guarantee the output tracking accuracy. The control voltage of the servo valve is shown in Figure 4.12. Because the control saturation is ± 10 V, the robust controller is designed to guarantee the dynamic tracking accuracy of the cylinder position under the control input constraint by the robust performance requirement (4.38).

4.4.4 Simulation and Experimental Result

The nonlinear simulation of Equations 4.1, 4.2, and 4.39 can be implemented by MATLAB/Hydraulic and Simechanics modules. The maximum stroke of the cylinder is 79 mm. So the range of the shoulder joint angle is from $-60°$ to $20°$ and the range of the elbow joint angle is from $40°$ to $130°$. Furthermore,

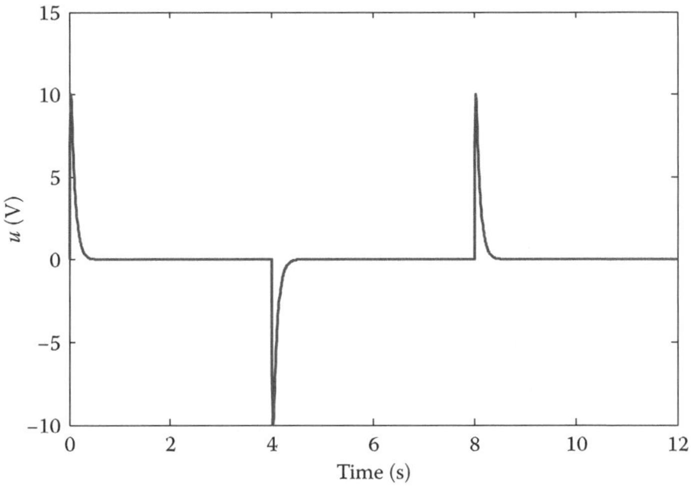

FIGURE 4.12
Designed controller result in time domain.

the relative variations δ_{V_t}, δ_b, $\delta_{K_{FLu}}$, and $\delta_{K_{FLf}}$ are all selected as 1, which means the maximum uncertainties. Then, two simulation results about the square response and the sinusoidal response of the maximum stroke are shown in Figures 4.13 through 4.16.

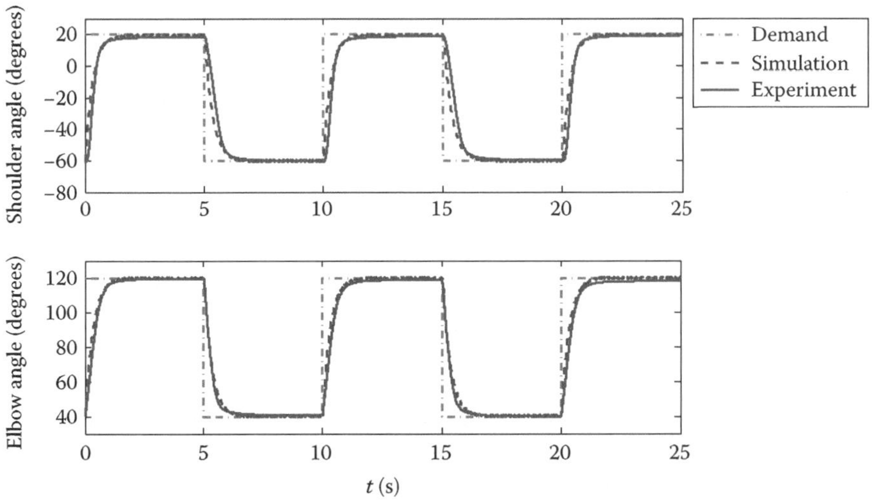

FIGURE 4.13
Square response of two joint angles in simulation and experiment.

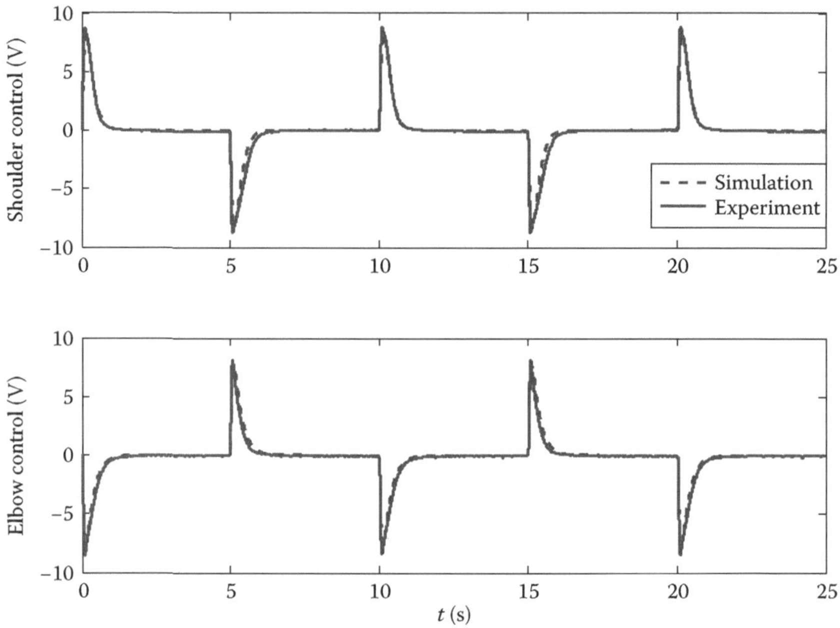

FIGURE 4.14
Control voltage of two servo valves in simulation and experiment for square demand.

The two square frequencies are 0.1 and 0.1 Hz and the sinusoidal frequencies are 0.3 and 0.5 Hz, respectively. Since the load on the shoulder joint is heavier than the elbow joint, the sinusoidal frequency for the shoulder motion is not allowed too high. The steady errors of two joint angles are less than 2, 0.5, respectively, shown in Figure 4.13, which means that the relative position errors are less than 5%. Owing to the small dynamic mechanical load on the elbow arm, the tracking accuracy is higher than the shoulder arm. The control voltages of two servo valves are no more than the control saturation ± 10 V as shown in Figure 4.14.

Similarly for the sinusoidal response, the maximum dynamic errors of two joint angles are less than 3°, 3°, respectively, shown in Figure 4.15. These steady errors are no more than the relative error 5% of demand input. The control voltages are also no more than ± 10 V as shown in Figure 4.16. So the two robust controllers of two hydraulic actuators can be validated effectively in simulation.

In this experiment, the supply pressure p_s is 40 bar, and the motor revolution is fixed at 1000 rpm. The joint angle is measured by an encoder and the control voltage is measured by a serial port terminal of the servo valve. The experiment results are very close to the corresponding simulation results as shown in Figures 4.13 through 4.16. The steady errors of two joint angles are less than 2°, 2°, respectively, in square response results. The maximum

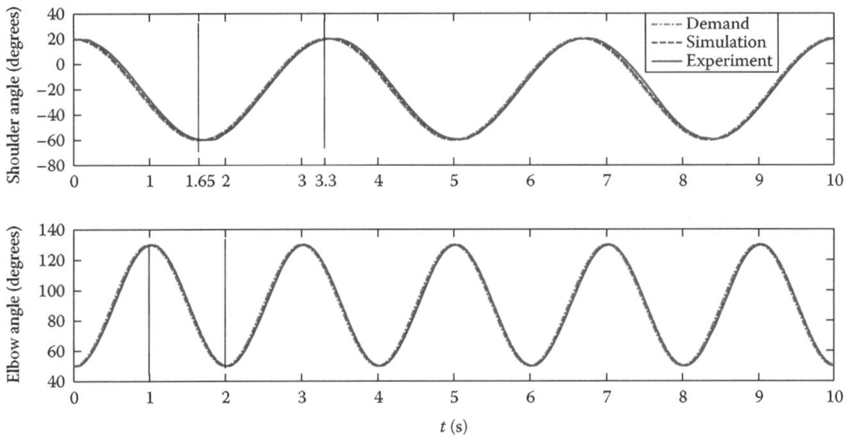

FIGURE 4.15
Sinusoidal response of two joint angles in simulation and experiment.

dynamic errors of two joint angles are less than 4°, 4°, respectively, in sinusoidal response results, which are no more than the relative error 5% of demand input. So the two robust controllers can guarantee the tracking position accuracy and dynamic response performance in the case of the robust stability. Furthermore, the control voltages of two servo valves are also no

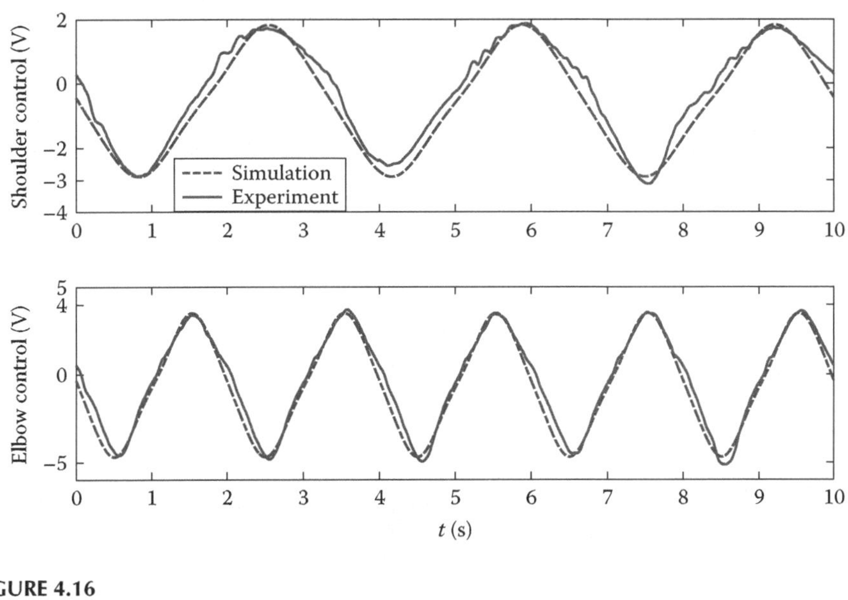

FIGURE 4.16
Control voltage of two servo valves in simulation and experiment for sinusoidal demand.

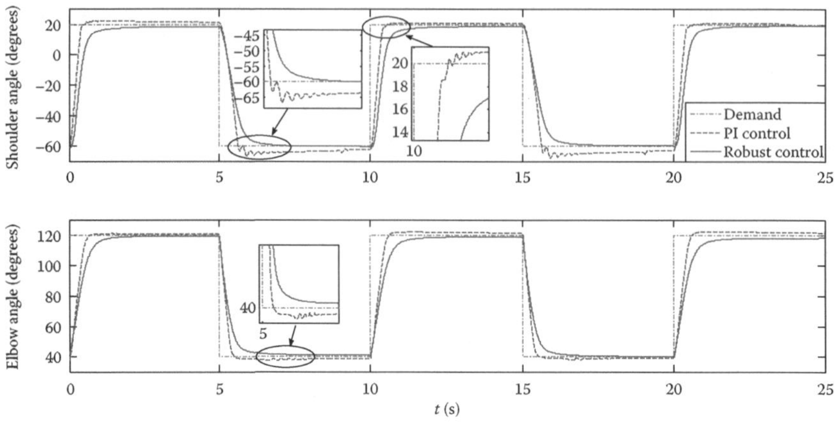

FIGURE 4.17
Square response of the experiment result by the two control methods.

more than the control saturation, which is reflected in the robust performance requirement e_u as shown in Equation 4.38.

The traditional PI control methods can also be used in this EHS. The experiment result comparison for the two control methods is shown in Figures 4.17 and 4.18. In Figure 4.17, although the step response by PI control is faster than by robust control, the transient chatter emerges in two responses of the shoulder, especially in the retraction process of the shoulder actuator. This chatter phenomenon is the main cause by the different hydraulic parameters, rapid changes of load forces between actuator extension and retraction. It shows that the robust H_∞ control method can eliminate the transient chatter

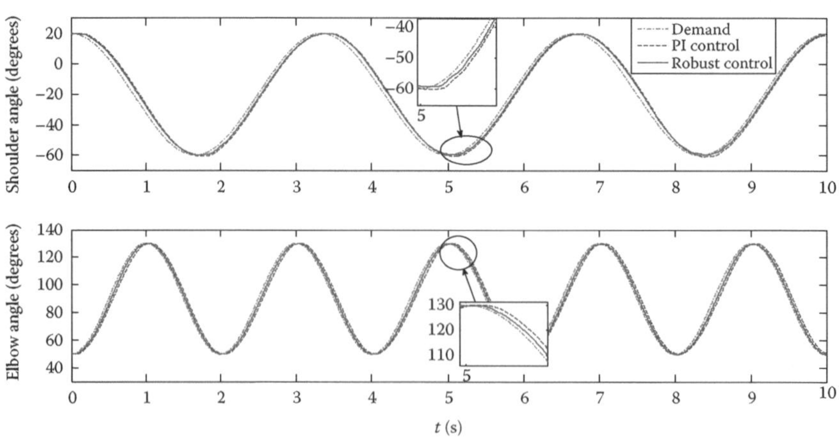

FIGURE 4.18
Sinusoidal response of the experiment result by the two control methods.

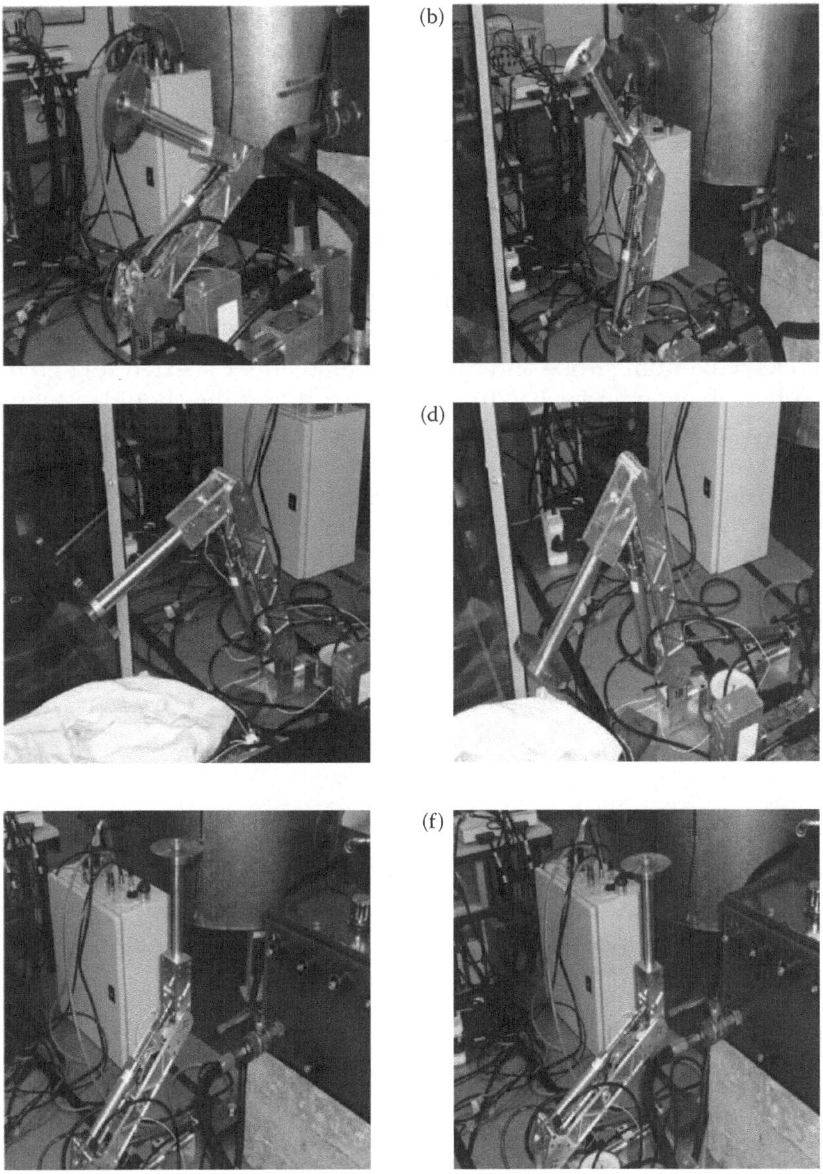

FIGURE 4.19
Snapshots of the sinusoidal experiment process. (a) Two cylinders retracted entirely ($t = 0$ s). (b) The elbow cylinder extended to the maximum stroke ($t = 1$ s). (c) The shoulder cylinder extended to the maximum stroke ($t = 1.65$ s). (d) The elbow cylinder retracted entirely ($t = 2$ s). (e) The elbow cylinder extended to the maximum stroke once again ($t = 3$ s). (f) The shoulder cylinder extended to the maximum stroke once again ($t = 3.3$ s).

problem caused by the structural disturbances and parametric uncertainties. The maximum steady error is 5° of the shoulder step response by PI control, which is a little larger than by robust control. The steady error of the elbow step response is 2°, less than that of the shoulder step because of the smaller load forces. In Figure 4.18, the hysteresis of sinusoidal response by PI control is more clearer than by robust control. It shows that the dynamic tracking performance is obviously improved by robust control method.

The video sequences of sinusoidal experiment process are as shown in Figure 4.19. The two-DOF robotic arm is controlled steady by two robust controllers with 1 kg disk load. Double cylinders are all retracted at initial time $t = 0$ s. Approximately 1 s later, the elbow cylinder is extended to reach its maximum stroke and the shoulder cylinder is extending gradually. Then the elbow cylinder has turned to retract and the shoulder cylinder is extended to reach its maximum stroke at $t = 1.65$ s. At $t = 2$ s, the elbow cylinder is retracted entirely and the shoulder cylinder turns to retract gradually. Then the elbow cylinder is extended to reach its maximum stroke once again at $t = 3$ s, which means the duration of the elbow joint motion has ended. But the shoulder cylinder is retracting gradually at this time. After 3.3 s, the shoulder cylinder is retracted entirely, which means the duration of the shoulder joint motion has ended.

5

Output Feedback Control Method

CONTENTS

In this chapter, an output feedback controller with a full-state observer is presented for a single-rod hydraulic actuator [19]. The system states are not measured except the cylinder displacement. Although this system is not a strict feedback system, the proposed controller theoretically guarantees the stability and convergence of EHS. In addition, the stable full-state observer is designed to estimate some unknown hydraulic states, which can be used in the backstepping controller. To verify this proposed backstepping control based on the full-state observer, an experiment is carried out. Experiment results are obtained from the hydraulic position control of the two-DOF robotic arm motion. Experiment results highlight the better dynamic performance of the proposed method in comparison to the conventional PI control method in some specific critical condition with high response frequency and large unknown external load.

5.1 Output Feedback Control Model of EHS

The electro-hydraulic system shown in Figure 2.1b comprises a servo valve, a single-rod cylinder, a fixed-displacement pump, a relief valve, and a variable load. The variable load Pl drives the two-DOF robotic arm manufactured by Italian Institute of Technology, which is referred as the Robotic BigDog. In this section, the nonlinear dynamic model of the single-rod hydraulic actuator will be discussed.

The dynamics of the servo valve is adopted by a second-order linear model mentioned in Equation 2.9 as follows:

$$\ddot{x}_v + 2\zeta_{sv}\omega_{sv}\dot{x}_v + \omega_{sv}^2 x_v = K_{sv}\omega_{sv}^2 u, \tag{5.1}$$

where x_v is the spool position of the servo valve, ζ_{sv} is the damping ratio, ω_{sv} is the natural frequency, and K_{sv} is the gain of control voltage u.

According to Equation 2.6, the flow equations of the single-rod cylinder can be described as follows:

$$Q_a = \begin{cases} C_d w x_v \sqrt{2(p_s - p_a)/\rho} & x_v \geq 0 \\ C_d w x_v \sqrt{2(p_a - p_r)/\rho} & x_v < 0 \end{cases},$$

$$Q_b = \begin{cases} C_d w x_v \sqrt{2(p_b - p_r)}/\rho & x_v \geq 0 \\ C_d w x_v \sqrt{2(p_s - p_b)}/\rho & x_v < 0 \end{cases} \tag{5.2}$$

where Q_a is the main load flow as $x_v \geq 0$, Q_b is the main load flow as $x_v < 0$, p_a and p_b are the pressure inside the two chambers of the cylinder, p_s is the supply pressure of the pump, C_d is the discharge coefficient, w is the area gradient of the servo valve spool, and ρ is the density of the hydraulic oil.

The two load flows Q_a and Q_b are handled like the following uniform form:

$$\begin{cases} Q_a = \left(\dfrac{1+\mathrm{sgn}(x_v)}{2}\right) C_d w x_v \sqrt{\dfrac{2}{\rho}(p_s - p_a)} \\ \qquad + \left(\dfrac{1-\mathrm{sgn}(x_v)}{2}\right) C_d w x_v \sqrt{\dfrac{2}{\rho}(p_a - p_r)} \\ Q_b = \left(\dfrac{1+\mathrm{sgn}(x_v)}{2}\right) C_d w x_v \sqrt{\dfrac{2}{\rho}(p_b - p_r)} \\ \qquad + \left(\dfrac{1-\mathrm{sgn}(x_v)}{2}\right) C_d w x_v \sqrt{\dfrac{2}{\rho}(p_s - p_b)} \end{cases}, \tag{5.3}$$

where the sgn(.) function is defined as

$$\mathrm{sgn}(x) = \begin{cases} -1 & x > 0 \\ 0 & x = 0 \\ 1 & x < 0 \end{cases}. \tag{5.4}$$

In Equation 5.3, the function $\mathrm{sgn}(x_v)$ should be smoothed in the derivation of the backstepping control. So $\mathrm{sgn}(x_v)$ can be replaced with $\tanh(kx_v)$ as follows [16]:

$$\mathrm{sgn}(x_v) \approx \tanh(kx_v) = \frac{e^{kx_v} - e^{-kx_v}}{e^{kx_v} + e^{-kx_v}}, \quad k \gg 0. \tag{5.5}$$

According to Equation 2.3, the flow-pressure continuous model of the hydraulic cylinder is shown as follows:

$$\begin{cases} A_a\dot{y} + C_{tl}(p_a - p_b) + (V_{0a} + A_a y)\dot{p}_a/\beta_e = Q_a \\ A_b\dot{y} + C_{tl}(p_a - p_b) - (V_{0b} - A_b y)\dot{p}_b/\beta_e = Q_b \end{cases}, \tag{5.6}$$

where y is the displacement of the piston, C_{tl} is the coefficient of the total leakage of the cylinder, β_e is the effective bulk modulus, A_a and A_b are the ram areas of the two chambers, and V_{0a} and V_{0b} are the initial total control volumes of the two cylinder chambers, respectively.

From Equation 2.2, the mechanical dynamic equation is shown as follows:

$$m\ddot{y} = p_L A_p - Ky - b\dot{y} - F_L, \tag{5.7}$$

where m is the load mass, K is the load spring constant, b is the viscous damping coefficient, F_f is the stick-slip friction caused by the viscous damp of oil, and F_L is the external load on the hydraulic actuator from the mechanical structure of the two-DOF robotic arm.

The stick-slip friction often includes two items: a stick phase $F_{fstatic}$ occurs when the velocity is within a small critical velocity range and a slip friction F_{fslip}, which is the same as the Coulomb friction model [30]. Here, F_f is defined as

$$F_f(y, \dot{y}) = F_{fstatic} + b(1 + \delta q_b)\dot{y}, \quad -1 \le \delta \le 1, \tag{5.8}$$

where b is the viscous damping coefficient of oil, q_b is the maximum relative uncertainties of b, and δ is the relative variation.

According to Remark 2.2, the H_∞ norm of F_L is bounded by a constant Δ_{F_L} as follows:

$$\left\| F_L(y, \dot{y}, \ddot{y}) \right\|_\infty < \Delta_{F_L}. \tag{5.9}$$

If these vectors are defined: the state vector $X = [x_1, x_2, x_3, x_4, x_5, x_6]^T = [y, \dot{y}, p_a, p_b, x_v, \dot{x}_v]^T$, the control inputs $u(t)$, which is the voltage of the servo valve, the displacement of piston $y(t)$ indirectly measured by the encoder, the sixth-order nonlinear dynamic model of EHS is constructed as follows:

$$\begin{cases} \dot{x}_1 = x_2 \\ \dot{x}_2 = \dfrac{1}{m}(x_3A_a - x_4A_b - Kx_1 - F_{fstatic} - b(1+q_b)x_2 - \Delta_{F_L}) \\ \dot{x}_3 = -\dfrac{\beta_e A_a x_2}{V_{0a} + A_a x_1} - \dfrac{\beta_e C_{tl}}{V_{0a} + A_a x_1}(x_3 - x_4) \\ \qquad + \dfrac{\beta_e C_d w x_5}{V_{0a} + A_a x_1}\left(\begin{array}{l} \left(\dfrac{1+\tanh(kx_5)}{2}\right)\sqrt{\dfrac{2}{\rho}(p_s - x_3)} \\ + \left(\dfrac{1-\tanh(kx_5)}{2}\right)\sqrt{\dfrac{2}{\rho}(x_3 - p_r)} \end{array}\right) \\ \dot{x}_4 = \dfrac{\beta_e A_b x_2}{V_{0b} - A_b x_1} - \dfrac{\beta_e C_{tl}}{V_{0b} - A_b x_1}(x_3 - x_4) \\ \qquad - \dfrac{\beta_e C_d w x_5}{V_{0b} - A_b x_1}\left(\begin{array}{l} \left(\dfrac{1+\tanh(kx_5)}{2}\right)\sqrt{\dfrac{2}{\rho}(x_4 - p_r)} \\ + \left(\dfrac{1-\tanh(kx_5)}{2}\right)\sqrt{\dfrac{2}{\rho}(p_s - x_4)} \end{array}\right) \\ \dot{x}_5 = x_6 \\ \dot{x}_6 = -\omega_{sv}^2 x_5 - 2\zeta_{sv}\omega_{sv}x_6 + K_{sv}\omega_{sv}^2 u \end{cases} \quad . \tag{5.10}$$

5.2 State Observer Design

Owing to the unknown states except x_1, a high-gain state observer is designed to estimate the full state, which can be used in the design of the backstepping control. In this section, different from the previous state observer in Reference 16, a stable observer is constructed by more processing steps. Then the global asymptotic convergence for the full-state observer is proved.

5.2.1 Full-State Observer Construction

According to Equation 5.10, the nonlinear model of the EHS can be rewritten as

$$\begin{aligned} \dot{X} &= AX + \phi(X) + Bu, \\ y &= CX, \end{aligned} \tag{5.11}$$

where

$$A = \begin{bmatrix} 0 & 1 & 0 & 0 & 0 & 0 \\ 0 & 0 & \frac{A_a}{m} & -\frac{A_b}{m} & 0 & 0 \\ 0 & 0 & -\frac{\beta_e C_{tl}}{V_{0a} + A_a x_1} & -\frac{\beta_e C_{tl}}{V_{0a} + A_a x_1} & 0 & 0 \\ 0 & 0 & -\frac{\beta_e C_{tl}}{V_{0b} - A_b x_1} & -\frac{\beta_e C_{tl}}{V_{0b} - A_b x_1} & 0 & 0 \\ 0 & 0 & 0 & 0 & 0 & 1 \\ 0 & 0 & 0 & 0 & -\omega_v^2 & -2\zeta_v\omega_v \end{bmatrix}, \tag{5.12}$$

$$B = \begin{bmatrix} 0 & 0 & 0 & 0 & 0 & K_{sv}\omega_v^2 \end{bmatrix}^T, \tag{5.13}$$

$$C = \begin{bmatrix} 1 & 0 & 0 & 0 & 0 & 0 \end{bmatrix}, \tag{5.14}$$

$$\phi(X) = \begin{bmatrix} 0 \\ \frac{1}{m}(-Kx_1 - F_{fstatic} - bx_2 - bp_b x_2 - \Delta_{FL}) \\ \frac{\beta_e C_d w x_5}{V_{0a} + A_a x_1}\begin{pmatrix} \left(\frac{1 + \tanh(kx_5)}{2}\right)\sqrt{\frac{2}{\rho}(p_s - x_3)} + \\ \left(\frac{1 - \tanh(kx_5)}{2}\right)\sqrt{\frac{2}{\rho}(x_3 - p_r)} \end{pmatrix} - \frac{\beta_e A_a x_2}{V_{0a} + A_a x_1} \\ -\frac{\beta_e C_d w x_5}{V_{0b} - A_b x_1}\begin{pmatrix} \left(\frac{1 + \tanh(kx_5)}{2}\right)\sqrt{\frac{2}{\rho}(x_4 - p_r)} + \\ \left(\frac{1 - \tanh(kx_5)}{2}\right)\sqrt{\frac{2}{\rho}(p_s - x_4)} \end{pmatrix} + \frac{\beta_e A_b x_2}{V_{0b} - A_b x_1} \\ 0 \\ 0 \end{bmatrix}. \tag{5.15}$$

Equivalent transformation is used in the third and fourth columns of matrix A; the new matrix A_0 can be defined as

$$A_0 \overset{\langle 3\rangle\times(-1)+\langle 4\rangle}{=} \begin{bmatrix} 0 & 1 & 0 & 0 & 0 & 0 \\ -\frac{K}{m} & 0 & \frac{A_a}{m} & -\frac{A_b}{m}-\frac{A_a}{m} & 0 & 0 \\ 0 & 0 & -\frac{\beta_e C_{tl}}{V_{0a}+A_a x_1} & 0 & 0 & 0 \\ 0 & 0 & -\frac{\beta_e C_{tl}}{V_{0b}-A_b x_1} & 0 & 0 & 0 \\ 0 & 0 & 0 & 0 & 0 & 1 \\ 0 & 0 & 0 & 0 & -\omega_v^2 & -2\zeta_v\omega_v \end{bmatrix}, \tag{5.16}$$

Since the leakage coefficient C_{tl} has a small value and hence little impact, two items $A_0(3,3)$ and $A_0(4,3)$ are neglected in this chapter. Hence, $A_0(3,3) = A_0(4,3) = 0$. To ensure the validation and convergence of the observer, two items $A_0(3,5)$ and $A_0(4,5)$ are replaced by other nonzero items as follows:

$$A_0(3,5) = \frac{\beta_e}{V_{0a}+A_a x_1} C_d w \sqrt{\frac{2}{\rho} p_s}, \tag{5.17}$$

$$A_0(4,5) = -\frac{\beta_e}{V_{0b}-A_b x_1} C_d w \sqrt{\frac{2\lambda}{\rho} p_s}, \tag{5.18}$$

where $0 < \lambda < 1$ is a constant.

Then, two elements $A_0(6,5)$ and $A_0(6,6)$ are moved into $\phi(X)$, which means $A_0(6,5) = A_0(6,6) = 0$. So the new matrix A_0 and $\phi(X)$ are defined as

$$A_1 = \begin{bmatrix} 0 & 1 & 0 & 0 & 0 & 0 \\ 0 & 0 & a_{23} & -a_{23}-a_{24} & 0 & 0 \\ 0 & 0 & 0 & 0 & a_{35} & 0 \\ 0 & 0 & 0 & 0 & -a_{45} & 0 \\ 0 & 0 & 0 & 0 & 0 & 1 \\ 0 & 0 & 0 & 0 & 0 & 0 \end{bmatrix}, \tag{5.19}$$

$$\phi_1(X) = \begin{bmatrix} 0 \\ \frac{1}{m}(-Kx_1 - F_{fstatic} - bx_2 - bp_b x_2 - \Delta_{FL}) \\ \frac{\beta_e C_d w x_5}{V_{0a} + A_a x_1}\left(\begin{array}{c}\left(\frac{1+\tanh(kx_5)}{2}\right)\sqrt{\frac{2}{\rho}(p_s - x_3)} \\ +\left(\frac{1-\tanh(kx_5)}{2}\right)\sqrt{\frac{2}{\rho}(x_3 - p_r)} - C_d w\sqrt{\frac{2}{\rho}p_s}\end{array}\right) \\ -\frac{\beta_e A_a x_2}{V_{0a} + A_a x_1} \\ -\frac{\beta_e C_d w x_5}{V_{0b} - A_b x_1}\left(\begin{array}{c}\left(\frac{1+\tanh(kx_5)}{2}\right)\sqrt{\frac{2}{\rho}(x_4 - p_r)} \\ +\left(\frac{1-\tanh(kx_5)}{2}\right)\sqrt{\frac{2}{\rho}(p_s - x_4)} - C_d w\sqrt{\frac{2\lambda}{\rho}p_s}\end{array}\right) \\ +\frac{\beta_e A_b x_2}{V_{0b} - A_b x_1} \\ 0 \\ -\omega_v^2 x_5 - 2\zeta_v \omega_v x_6 \end{bmatrix}, \tag{5.20}$$

where

$$a_{23} = \frac{A_a}{m}, \tag{5.21}$$

$$a_{24} = \frac{A_b}{m}, \tag{5.22}$$

$$a_{35} = \frac{\beta_e}{V_{0a} + A_a x_1} C_d w \sqrt{\frac{2}{\rho} p_s}, \tag{5.23}$$

$$a_{45} = \frac{\beta_e}{V_{0b} - A_b x_1} C_d w \sqrt{\frac{2\lambda}{\rho} p_s}. \tag{5.24}$$

To estimate X, the high-gain observer is designed as follows:

$$\begin{aligned} \dot{\hat{X}} &= A_1 \hat{X} + \phi_1(\hat{X}) + Bu + \Gamma(y - \hat{y}), \\ \Gamma &= \begin{bmatrix}\Gamma_1 & \Gamma_2 & \Gamma_3 & \Gamma_4 & \Gamma_5 & \Gamma_6\end{bmatrix}^T, \quad \Gamma_i > 0, \quad i = 1, \ldots, 6. \end{aligned} \tag{5.25}$$

where Γ is the high gain of the observer, y is the measured value of displacement, $\hat{y} = \hat{x}_1$.

5.2.2 Observer Convergence

If the designed observer is fast convergence, the next estimated state $\hat{X}(k+1)$ can be computed by the current estimated state $\hat{X}(k)$ and current control $u(k)$ according to Equation 5.25. Now the stability and convergence of this observer are analyzed. Since Equation 5.11 is subtracted by Equation 5.25, the error state model of the observer is defined as

$$\dot{\tilde{X}} = A_c\tilde{X} + \delta_\phi(X, \tilde{X}), \tag{5.26}$$

where

$$\begin{aligned}\tilde{X} &= \begin{bmatrix}\tilde{x}_1 & \tilde{x}_2 & \tilde{x}_3 & \tilde{x}_4 & \tilde{x}_5 & \tilde{x}_6\end{bmatrix}^T \\ &= \begin{bmatrix}x_1-\hat{x}_1 & x_2-\hat{x}_2 & x_3-\hat{x}_3 & x_4-\hat{x}_4 & x_5-\hat{x}_5 & x_6-\hat{x}_6\end{bmatrix}^T,\end{aligned} \tag{5.27}$$

$$A_c = A_1 - LC = \begin{bmatrix}-\Gamma_1 & 1 & 0 & 0 & 0 & 0\\ -\Gamma_2 & 0 & a_{23} & -a_{23}-a_{24} & 0 & 0\\ -\Gamma_3 & 0 & 0 & 0 & a_{35} & 0\\ -\Gamma_4 & 0 & 0 & 0 & -a_{45} & 0\\ -\Gamma_5 & 0 & 0 & 0 & 0 & 1\\ -\Gamma_6 & 0 & 0 & 0 & 0 & 0\end{bmatrix}, \tag{5.28}$$

$$\delta_\phi(X, \tilde{X}) = \phi_1(X) - \phi_1(\hat{X}). \tag{5.29}$$

The characteristic polynomial of error system matrix A_c is described as

$$\begin{aligned}\det|sI - A_c| = {} & s^6 + \Gamma_1 s^5 + \Gamma_2 s^4 + [a_{23}\Gamma_3 + (a_{23}+a_{24})\Gamma_4]s^3 \\ & + [a_{23}a_{35}\Gamma_5 - (a_{23}+a_{24})a_{45}\Gamma_5]s^2 \\ & + [a_{23}a_{35}\Gamma_6 - (a_{23}+a_{24})a_{45}\Gamma_6]s = 0.\end{aligned} \tag{5.30}$$

Theorem 5.1

If the high gain of the observer Γ is satisfied to the following condition (5.31), then the matrix A_c is Hurwitz after removing the only one eigenvalue 0. In other words, all the eigenvalues of A_c has a negative real part.

$$\begin{cases} \Gamma_1 > 0 \\ \Gamma_2 > 0 \\ a_{23}\Gamma_3 + (a_{23} + a_{24})\Gamma_4 > 0 \\ a_{23}a_{35}\Gamma_5 - (a_{23} + a_{24})a_{45}\Gamma_5 > 0 \\ a_{23}a_{35}\Gamma_6 - (a_{23} + a_{24})a_{45}\Gamma_6 > 0 \\ \Gamma_1\Gamma_2 - a_{23}\Gamma_3 - (a_{23} + a_{24})\Gamma_4 > 0 \\ \Gamma_1\Gamma_2[a_{23}\Gamma_3 + (a_{23} + a_{24})\Gamma_4] - [a_{23}\Gamma_3 + (a_{23} + a_{24})\Gamma_4]^2 \\ \quad - \Gamma_1^2\Gamma_5[a_{23}a_{35} - (a_{23} + a_{24})a_{45}] + \Gamma_1\Gamma_6[a_{23}a_{35} - (a_{23} + a_{24})a_{45}] > 0 \end{cases} . \tag{5.31}$$

Proof. Equation 5.30 shows that one eigenvalue of the matrix A_c is 0 easily. After removing the eigenvalue 0, Equation 5.30 can be simplified as

$$s^5 + c_4s^4 + c_3s^3 + c_2s^2 + c_1s + c_0 = 0. \tag{5.32}$$

According to Routh criterion, the necessary and sufficient conditions that all characteristic roots have a negative real part must satisfy two properties as follows:

I. All coefficients are greater than zero, which means $c_i > 0 (i = 0, \ldots, 4)$.
II. Arrange the coefficients c_i in the following form:

$$\begin{array}{c|ccc|} s^5 & 1 & c_3 & c_1 \\ s^4 & c_4 & c_2 & c_0 \\ s^3 & \dfrac{c_4c_3 - c_2}{c_4} & \dfrac{c_4c_1 - c_0}{c_4} & 0 \\ s^2 & c_2 - \dfrac{c_4^2c_1 - c_4c_0}{c_4c_3 - c_2} & c_0 & 0 \\ s^1 & \dfrac{c_4c_1 - c_0}{c_4} - \dfrac{\dfrac{c_4c_0c_3 - c_2c_0}{c_4}}{c_2 - \dfrac{c_4^2c_1 - c_4c_0}{c_4c_3 - c_2}} & 0 & 0 \\ s^0 & c_0 & 0 & 0 \end{array} , \tag{5.33}$$

where all the first worksheet of Equation 5.33 should be greater than 0.

According to property (II), if Equation 5.30 is substituted into the first worksheet of Equation 5.33, then the high gains of the observer $\Gamma_i (i = 1, \ldots, 6)$ are satisfied to five inequality conditions as shown in Equation 5.31. ■

According to the fourth and fifth inequalities of Equation 5.31, the following property is satisfied:

$$a_{23}a_{35} > (a_{23} + a_{24})a_{45}. \tag{5.34}$$

When Equations 5.23 and 5.24 are substituted into Equation 5.34, the constant parameter is satisfied as follows:

$$\lambda < \frac{A_a^2}{A_a^2 + A_b^2}. \tag{5.35}$$

Now, the dynamic estimation error in Equation 5.26 is considered. If a small positive constant ε is defined, then the new estimation error is described as

$$\begin{aligned} \tilde{\eta} &= \begin{bmatrix} \tilde{\eta}_1 & \tilde{\eta}_2 & \tilde{\eta}_3 & \tilde{\eta}_4 & \tilde{\eta}_5 & \tilde{\eta}_6 \end{bmatrix}^T \\ &= \begin{bmatrix} \frac{\tilde{x}_1}{\varepsilon^4} & \frac{\tilde{x}_2}{\varepsilon^3} & \frac{\tilde{x}_3}{\varepsilon^2} & \frac{\tilde{x}_4}{\varepsilon^2} & \frac{\tilde{x}_5}{\varepsilon} & \tilde{x}_6 \end{bmatrix}^T. \end{aligned} \tag{5.36}$$

After $\tilde{\eta}$ is substituted into Equation 5.26, the new error state equation is described as

$$\varepsilon\dot{\tilde{\eta}} = A_\eta\tilde{\eta} + \delta_\phi(\eta, \tilde{\eta}, \varepsilon), \tag{5.37}$$

where

$$A_\eta = \begin{bmatrix} -\Gamma_1/\varepsilon & 1 & 0 & 0 & 0 & 0 \\ -\Gamma_2/\varepsilon^2 & 0 & a_{23} & -a_{23} - a_{24} & 0 & 0 \\ -\Gamma_3/\varepsilon^3 & 0 & 0 & 0 & a_{35} & 0 \\ -\Gamma_4/\varepsilon^3 & 0 & 0 & 0 & -a_{45} & 0 \\ -\Gamma_5/\varepsilon^4 & 0 & 0 & 0 & 0 & 1 \\ -\Gamma_6/\varepsilon^5 & 0 & 0 & 0 & 0 & 0 \end{bmatrix}, \tag{5.38}$$

$$\delta_\phi(\eta, \hat{\eta}, \varepsilon) = \begin{bmatrix} 0 \\ \varepsilon\delta_2(\eta_2, \hat{\eta}_2, \varepsilon) \\ \delta_3(\eta_5, \hat{\eta}_5, \varepsilon) \\ \delta_4(\eta_5, \hat{\eta}_5, \varepsilon) \\ 0 \\ \varepsilon\delta_6(\eta_6, \hat{\eta}_6, \varepsilon) \end{bmatrix}. \tag{5.39}$$

Substituting Γ_i in Equation 5.28 with the first column of Equation 5.38, respectively, the two characteristic polynomials of matrix A_η and A_c are identical, which means the properties of all characteristic roots for A_η is equivalent to A_c. So, the matrix A_η is also Hurwitz after removing one eigenvalue 0 and the seven inequality conditions (5.31) are still available for the new error state Equation 5.37.

Theorem 5.2

If $\phi_1(X)$ is continuous and differentiable, and its first-order derivative is bounded, Equation 5.39 satisfies the following condition:

$$\delta_\phi(\eta, \tilde{\eta}, \varepsilon) \leq \varepsilon\mu_\delta \|\tilde{\eta}\|_2, \tag{5.40}$$

where μ_δ is a positive constant.

Proof. According to Equation 5.20, $\phi_1(X)$ is continuous and differentiable. Because all the states described specified physical variables, the first-order derivative of $\phi_1(X)$ is bounded. According to Equation 5.36, the first-order derivative of $\phi_1(X)$ is still bounded, although its form derives from $\phi_1(X)$.

Then Equation 5.39 can also be described as

$$\delta_\phi(\eta, \hat{\eta}, \varepsilon) = \phi_1(\eta) - \phi_1(\hat{\eta}) = \begin{bmatrix} 0 \\ \varepsilon\delta_2(\eta_2, \hat{\eta}_2, \varepsilon) \\ \varepsilon^2\delta_3(\eta_5, \hat{\eta}_5, \varepsilon) \\ \varepsilon^2\delta_4(\eta_5, \hat{\eta}_5, \varepsilon) \\ 0 \\ \varepsilon\delta_6(\eta_6, \hat{\eta}_6, \varepsilon) \end{bmatrix}. \tag{5.41}$$

According to Lagrange mean value theorem, if $\delta_i(\eta_i, \hat{\eta}_i, \varepsilon)(i = 2, 3, 4, 6)$ is continuous and differentiable, a point η_i^0 exists such that

$$\frac{\phi_{1i}(\eta_i) - \phi_{1i}(\hat{\eta}_i)}{\eta_i - \hat{\eta}_i} = \delta_i'(\eta_i^0), \quad \eta_i^0 \in (\eta_i, \hat{\eta}_i), \tag{5.42}$$

where $\delta_i'(\eta_i^0)$ is the derivative at η_i^0.

If η_i and $\hat{\eta}_i$ belong to M_i, which is a bounded neighborhood, η_i^0 also belongs to M_i and the following condition is satisfied:

$$\frac{\phi_{1i}(\eta_i) - \phi_{1i}(\hat{\eta}_i)}{\eta_i - \hat{\eta}_i} \leq \sup_{\eta_i^0 \in M_i} \left(\delta_i'(\eta_i^0)\right). \tag{5.43}$$

Thus, we can see that

$$\|\delta_\phi(\eta,\tilde{\eta},\varepsilon)\|_2 = \left\| \begin{bmatrix} 0 \\ \varepsilon\delta_2(\eta_2,\hat{\eta}_2,\varepsilon) \\ \varepsilon^2\delta_3(\eta_5,\hat{\eta}_5,\varepsilon) \\ \varepsilon^2\delta_4(\eta_5,\hat{\eta}_5,\varepsilon) \\ 0 \\ \varepsilon\delta_6(\eta_6,\hat{\eta}_6,\varepsilon) \end{bmatrix} \right\|_2 = \varepsilon \left\| \begin{bmatrix} 0 \\ \sup\limits_{\eta_2^0\in M_2} \left(\delta_2'(\eta_2^0\right)\tilde{\eta}_2 \\ \varepsilon \sup\limits_{\eta_3^0\in M_3} \left(\delta_3'(\eta_3^0\right)\tilde{\eta}_3 \\ \varepsilon \sup\limits_{\eta_4^0\in M_4} \left(\delta_4'(\eta_4^0\right)\tilde{\eta}_4 \\ 0 \\ \sup\limits_{\eta_6^0\in M_6} \left(\delta_6'(\eta_6^0\right)\tilde{\eta}_6 \end{bmatrix} \right\|_2$$

$$< \varepsilon \left\| \begin{bmatrix} 0 \\ \sup\limits_{\eta_2^0\in M_2} \left(\delta_2'(\eta_2^0\right)\tilde{\eta}_2 \\ \sup\limits_{\eta_3^0\in M_3} \left(\delta_3'(\eta_3^0\right)\tilde{\eta}_3 \\ \sup\limits_{\eta_4^0\in M_4} \left(\delta_4'(\eta_4^0\right)\tilde{\eta}_4 \\ 0 \\ \sup\limits_{\eta_6^0\in M_6} \left(\delta_6'(\eta_6^0\right)\tilde{\eta}_6 \end{bmatrix} \right\|_2 . \tag{5.44}$$

If a positive constant is satisfied, the following condition:

$$\mu_\delta = \max\left(\left| \sup_{\eta_i^0\in M_i} \left(\delta_i'(\eta_i^0\right) \right| \right), \tag{5.45}$$

and taking μ_δ into Equation 5.44, Equation 5.40 is satisfied. ■

Theorem 5.3

In Equation 5.37, if the positive constant ε satisfies the following condition (5.45), then the error vector $\tilde{\eta}$ converges to 0, which means $\eta \to \hat{\eta}$:

$$0 < \varepsilon < \frac{1}{2\lambda_{\max}(P_\eta)\mu_\delta}, \tag{5.46}$$

where P_η is a positive definite matrix and $\lambda_{\max}(P_\eta)$ is the maximum eigenvalues of matrix P_η.

Proof. A Lyapunov candidate function is defined as

$$V_\eta = \varepsilon\tilde{\eta}^T P_\eta \tilde{\eta}. \tag{5.47}$$

Since A_η is a Hurwitz matrix, a positive definite matrix P_η exists such that

$$A_\eta^T P_\eta + P_\eta A_\eta = -\begin{bmatrix} I_{5\times 5} & 0 \\ 0 & 0 \end{bmatrix}. \tag{5.48}$$

If Equation 5.37 is combined with Equations 5.47 and 5.48, then the derivative of V_η is

$$\dot{V}_\eta = -\tilde{\eta}^T \tilde{\eta} + 2\tilde{\eta}^T P_\eta \delta_\phi(\eta, \hat{\eta}, \varepsilon). \tag{5.49}$$

So, $\dot{V}_\eta$ satisfies the following inequality:

$$\begin{aligned} \dot{V}_\eta &\leq -\|\tilde{\eta}\|_2^2 + 2\tilde{\eta}^T P_\eta \delta_\phi(\eta, \hat{\eta}, \varepsilon) \\ &\leq -\|\tilde{\eta}\|_2^2 + 2\varepsilon\lambda_{\max}(P_\eta)\mu_\delta \|\tilde{\eta}\|_2^2 \\ &\leq -(1 - 2\varepsilon\lambda_{\max}(P_\eta)\mu_\delta) \|\tilde{\eta}\|_2^2. \end{aligned} \tag{5.50}$$

If ε satisfies the condition shown in Equation 5.46, then $\dot{V}_\eta < 0$. Therefore, the error vector $\tilde{\eta}$ converges to 0 and the system described by Equation 5.37 is exponentially stable. ■

Since $\eta \to \hat{\eta}$, considering the relationship between η and X in Equation 5.36, we see that $X \to \hat{X}$.

5.3 Nonlinear Backstepping Control

The dynamic model of EHS in Equation 5.10 shows that the system order is six. But the relative degree is only five. Therefore, this EHS is not a strict feedback nonlinear system and then the recursive procedure of the backstepping control method cannot be used in Equation 5.10 directly. In this section, the dynamic model is divided into two parts to design the backstepping controller after a simple transformation of the pressure error from two chambers.

5.3.1 Backstepping Procedure

Equation 5.10 is considered to design the backstepping controller by recursive procedure. At first, a pressure error is defined as

$$
\begin{aligned}
e_p &= p - p_d, \\
p &= \frac{A_a x_3 - A_b x_4}{A_a}, \\
p_d(t) &= \frac{K x_{1d}(t)}{A_a} + \frac{F_{fstatic}}{A_a} + \frac{\Delta F_L}{A_a},
\end{aligned}
\tag{5.51}
$$

where p_d is the virtual command of the load pressure and p is the actual load pressure, and x_{1d} is the displacement command of the system input.

The error of x_5 is defined as

$$e_5 = x_5 - x_{5d}, \tag{5.52}$$

and the virtual command x_{5d} is satisfied as

$$x_{5d} = -\frac{1}{g_2}(k_1 e_p - \dot{p}_d + g_1), \tag{5.53}$$

where k_1 is a positive constant, the functions g_1 and g_2 are

$$
\begin{aligned}
g_1 = &-\frac{\beta_e A_a x_2}{V_{0a} + A_a x_1} - \frac{\beta_e C_{tl}}{V_{0a} + A_a x_1}(x_3 - x_4) - \frac{\beta_e A_b^2 x_2}{A_a(V_{0b} - A_b x_1)} \\
&+ \frac{\beta_e A_b C_{tl}}{A_a(V_{0b} - A_b x_1)}(x_3 - x_4),
\end{aligned}
\tag{5.54}
$$

$$
\begin{aligned}
g_2 = &\frac{\beta_e}{V_{0a} + A_a x_1}\left(\begin{aligned}&\left(\frac{e^{kx_5}}{e^{kx_5} + e^{-kx_5}}\right) C_d w \sqrt{\frac{2}{\rho}(p_s - x_3)} \\ &+ \left(\frac{e^{-kx_5}}{e^{kx_5} + e^{-kx_5}}\right) C_d w x_5 \sqrt{\frac{2}{\rho}(x_3 - p_r)}\end{aligned}\right) \\
&+ \frac{\beta_e A_b}{A_a(V_{0b} - A_b x_1)}\left(\begin{aligned}&\left(\frac{e^{kx_5}}{e^{kx_5} + e^{-kx_5}}\right) C_d w \sqrt{\frac{2}{\rho}(x_4 - p_r)} \\ &+ \left(\frac{e^{-kx_5}}{e^{kx_5} + e^{-kx_5}}\right) C_d w x_5 \sqrt{\frac{2}{\rho}(p_s - x_4)}\end{aligned}\right).
\end{aligned}
\tag{5.55}
$$

Then the Lyapunov candidate function V_3 is defined as

$$V_3 = \frac{1}{2} e_p^2. \tag{5.56}$$

Since the derivative of V_3 is

$$\dot{V}_3 = e_p \dot{e}_p = e_p \left(\frac{A_a \dot{x}_3 - A_b \dot{x}_4}{A_a} - \dot{p}_d \right) = e_p (g_2 x_5 + g_1 - \dot{p}_d), \tag{5.57}$$

we can see that

$$\begin{aligned} & -k_1 e_p + g_2 e_5 \\ &= -k_1 e_p + g_2 (x_5 - x_{5d}) \\ &= -k_1 e_p + g_2 \left(x_5 + \frac{1}{g_2} (k_1 e_p - \dot{p}_d + g_1) \right) \\ &= g_2 x_5 + g_1 - \dot{p}_d. \end{aligned} \tag{5.58}$$

After Equation 5.57 is substituted into Equation 5.56, the derivative of V_3 is described as

$$\dot{V}_3 = -k_1 e_p^2 + g_2 e_p e_5. \tag{5.59}$$

The function V_3 is expanded and the Lyapunov candidate function V_4 is defined as

$$V_4 = V_3 + \frac{1}{2k_2} e_5^2, \tag{5.60}$$

where k_2 is a positive constant.

Then the derivative of V_4 is

$$\begin{aligned} \dot{V}_4 &= \dot{V}_3 + \frac{e_5}{k_2} \dot{e}_5 \\ &= -k_1 e_p^2 + g_2 e_p e_5 + \frac{e_5}{k_2} (x_6 - \dot{x}_{5d}). \end{aligned} \tag{5.61}$$

If the virtual command x_{6d} is satisfied as

$$x_{6d} = -k_2 g_2 e_p + \dot{x}_{5d} - e_5, \tag{5.62}$$

then we see that

$$\dot{V}_4 = -k_1 e_p^2 - \frac{e_5^2}{k_2} + \frac{e_5 e_6}{k_2}, \tag{5.63}$$

where the error of x_6 is defined as

$$e_6 = x_6 - x_{6d}. \tag{5.64}$$

Similarly, V_4 is expanded and the Lyapunov candidate function V_5 is defined as

$$V_5 = V_4 + \frac{1}{2k_3} e_6^2, \tag{5.65}$$

where k_3 is a positive constant.

The derivative of V_5 is

$$\begin{aligned}\dot{V}_5 &= \dot{V}_4 + \frac{e_6}{k_3}\dot{e}_6 \\ &= -k_1 e_p^2 - \frac{e_5^2}{k_2} + \frac{e_5 e_6}{k_2} + \frac{e_6}{k_3}(\dot{x}_6 - \dot{x}_{6d}) \\ &= -k_1 e_p^2 - \frac{e_5^2}{k_2} + \frac{e_5 e_6}{k_2} + \frac{e_6}{k_3}(-\omega_{sv}^2 x_5 - 2\zeta_v \omega_{sv} x_6 + K_{sv}\omega_{sv}^2 u - \dot{x}_{6d}).\end{aligned} \tag{5.66}$$

If the controller u is designed as

$$u = \frac{1}{K_{sv}\omega_{sv}^2}\left(\omega_{sv}^2 x_5 + 2\zeta_{sv}\omega_{sv} x_6 + \dot{x}_{6d} - \frac{k_3}{k_2}e_5 - e_6\right), \tag{5.67}$$

then we see that

$$\dot{V}_5 = -k_1 e_p^2 - \frac{e_5^2}{k_2} - \frac{e_6^2}{k_3} < 0. \tag{5.68}$$

So far, the recursive procedure for the third-order equivalent model of EHS ends. By the construction of three Lyapunov functions, the stability of the variable states $x_i (i = 3, 4, 5, 6)$ is analyzed from the third equation to the sixth equation in Equation 5.10.

5.3.2 Controller Design with Observer

Owing to unknown variable states x_5 and x_6, the variables e_5, e_6, $\dot{x}_{6d}$ cannot be computed and the controller u shown in Equation 5.57 cannot be used in EHS directly. In this section, the backstepping controller is further processed by state estimation $\hat{x}_i (i = 1, \ldots, 6)$.

At first, the first and second equations in Equation 5.10 can be described as the second-order linear nonhomogeneous model as follows:

$$\ddot{x}_1 + \alpha_1 \dot{x}_1 + \alpha_2 x_1 = \frac{A_a p - F_{fstatic} - \Delta_{FL}}{m}, \tag{5.69}$$

where

$$\alpha_1 = \frac{b(1 + p_b)}{m}, \quad \alpha_2 = \frac{K}{m}. \tag{5.70}$$

Second, the virtual command of load pressure p_d is designed according to different forms of the displacement command x_{1d} as follows:

1. If $x_{1d} = \upsilon \sin \omega t$ (υ is the amplitude and ω is the angular frequency of the sinusoidal function), p_d is redesigned as

$$\begin{aligned} p_d(t) &= p_{ds}(t) + p_{dd}(t), \\ p_{ds}(t) &= \frac{K\upsilon \sin \omega t}{A_a} + \frac{\Delta F_L}{A_a}, \\ p_{dd}(t) &= -\sigma_1 x_2(t), \end{aligned} \tag{5.71}$$

where σ_1 is a positive constant. After the item p_{dd} is involved in p_d, the characteristics roots of model (5.69) can be reassigned to improve the dynamic performance of x_1.

Compared to the third equation in Equation 5.51, the stick item $F_{fstatic}$ is not included in p_d shown in the second equation in Equation 5.71 since the actual displacement x_1 will be regulated dynamically.

2. If x_{1d} is the step command, we consider the similar form

$$x_{1d} = \upsilon(1 - e^{-\tau t}), \tag{5.72}$$

where τ is an asymptotic time constant.

Then, p_d is redesigned as

$$\begin{aligned} p_d(t) &= p_{ds}(t) + p_{dd}(t), \\ p_{ds}(t) &= \frac{K\upsilon(1 - e^{-\tau t})}{A_a} + \frac{b\upsilon e^{-\tau t}}{A_a} + \frac{\Delta F_L}{A_a} + F_{fstatic}, \\ p_{dd}(t) &= -\sigma_1 x_2(t). \end{aligned} \tag{5.73}$$

After Equation 5.71 or 5.73 is substituted into Equation 5.69, the model can be described by Laplace transform as follows:

$$s^2 x_1(s) + (\alpha_1 + \sigma_1) s x_1(s) + \alpha_2 x_1(s) = p'(s), \tag{5.74}$$

where

$$p'(s) = \frac{A_a(p_{ds} + e_p) - F_{fstatic} - \Delta_{FL}}{m}. \tag{5.75}$$

If $p'(s)$ is considered as the system input, and $x_1(s)$ is the system output, since $\alpha_1 + \sigma_1 > 0, \alpha_2 > 0$, the two characteristics roots of system (5.74) have all negative real parts, which means $x_1(t)$ is asymptotically stable.

When $t \to \infty$, $s \to 0$ in Equation 5.74. According to the backstepping procedure in section A, the pressure error e_p converges to 0. Then, from Equation 5.74, we see that

$$x_1(t) \to \frac{A_a p_{ds} - F_{fstatic} - \Delta_{FL}}{\alpha_2 m}. \tag{5.76}$$

Now Equation 5.71 or 5.73 is substituted into Equation 5.76, we see that

$$x_1(t) \rightarrow \frac{Kx_{1d} + \Delta_{\max}(F_L) - \Delta_{FL}}{m} = \frac{Kx_{1d}(t)}{m\alpha_2} = x_{1d}. \tag{5.77}$$

The above equation shows that the system output $x_1(t)$ can be controlled to track its command x_{1d} by the equivalent system input $p'(s)$. The parameter σ_1 is designed to make x_1 fast converge to x_{1d}.

According to the Lyapunov analysis in section A, the pressure error e_p converges to 0 exponentially, which means the dynamic response of load pressure p can be satisfied to the system requirement by the regulation of parameter k_1. In the expression of $p_{ds}(t)$ shown in Equations 5.71 and 5.73, the dynamic external load error $F_L(y, \dot{y}, \ddot{y})$ and Δ_{F_L} exist. But this error has little impact on the convergence of x_1.

Finally, the actual states $x_i(i = 2, \ldots, 6)$ are replaced by their estimation and the backstepping controller can be described as

$$\begin{cases} \hat{u} = \dfrac{1}{K_{sv}\omega_{sv}^2}\left(\omega_{sv}^2\hat{x}_5 + 2\zeta_{sv}\omega_{sv}\hat{x}_6 + \dot{\hat{x}}_{6d} - \dfrac{k_3}{k_2}\hat{e}_5 - \hat{e}_6\right) \\ \hat{e}_p = \dfrac{A_a\hat{x}_3 - A_b\hat{x}_4}{A_a} - p_d \\ \hat{x}_{5d} = -\dfrac{1}{g_2}(k_1\hat{e}_p - \dot{p}_d + g_1) \\ \hat{x}_{6d} = -k_2 g_2\hat{e}_p + \dot{\hat{x}}_{5d} - \hat{e}_5 \\ \hat{e}_5 = \hat{x}_5 - \hat{x}_{5d} \\ \hat{e}_6 = \hat{x}_6 - \hat{x}_{6d} \end{cases}, \tag{5.78}$$

where $\hat{x}_i(i = 3, 4, 5, 6)$ can be recursively computed by the observer Equation 5.25.

5.3.3 Stability Discussion of EHS

Now the stability of EHS should be analyzed after the full-state feedback control is realized. In Equation 5.11, the controller u is replaced by $\hat{u}$ as follows:

$$\dot{X} = AX + \phi(x) + B\hat{u}. \tag{5.79}$$

Combined with Equations 5.10, 5.52, 5.53, 5.62, 5.64, and 5.67, the three state errors satisfy the following conditions:

$$\begin{aligned}
\dot{e}_p &= -k_1 e_p + g_2 e_5, \\
\dot{e}_5 &= -e_5 - k_2 g_2 e_p + e_6, \\
\dot{e}_6 &= -\frac{k_3}{k_2} e_5 - e_6.
\end{aligned} \tag{5.80}$$

According to Equations 5.74 through 5.76, the state error x_1 equation is satisfied:

$$\ddot{e}_1(s) + (\alpha_1 + \sigma_1)\dot{e}_1(s) + \alpha_2 e_1(s) = \frac{A_a e_p}{m}. \tag{5.81}$$

If the full-state error is defined as

$$\begin{aligned}
Z &= \begin{bmatrix} z_1 & z_2 & z_3 & z_4 & z_5 \end{bmatrix}^T \\
&= \begin{bmatrix} e_1 & e_2 & e_p & e_5 & e_6 \end{bmatrix}^T,
\end{aligned} \tag{5.82}$$

and combined with Equations 5.11, 5.79 through 5.81, we see that

$$\begin{aligned}
\dot{Z} &= A_e Z + B_e Z + B(u - \hat{u}) \\
&= A_e Z + B_e Z + B\tilde{u}(X, \tilde{X}, x_{1d}),
\end{aligned} \tag{5.83}$$

where

$$A_e = \begin{bmatrix} -1 & & & & \\ -\alpha_2 & -(\alpha_1 + \sigma_1) & & & \\ & & -k_1 & & \\ & & & -1 & \\ & & & & -1 \end{bmatrix}, \tag{5.84}$$

$$B_e = \begin{bmatrix} 0 & \frac{A_a}{m} e_p & g_2 e_5 & -k_2 g_2 e_p + e_6 & -\frac{k_3}{k_2} e_5 \end{bmatrix}^T, \tag{5.85}$$

$$\tilde{u}(X, \tilde{X}, x_{1d}) = u(X, x_{1d}) - \hat{u}(\hat{X}, x_{1d}). \tag{5.86}$$

The observer error dynamics shown in Equation 5.37 is fast dynamics adopted by the high-gain state observer. On the contrary, the tracking error dynamics shown in Equation 5.83 is slow dynamics. Therefore, according to Equation 5.25, the gain of the observer Γ and the estimation of controller $\hat{u}$ can guarantee that the state estimation $\hat{X}$ converge to X quickly, and then $\hat{u} \to u$. This explanation is simply described as

$$\underset{\dot{X}}{\underset{\downarrow}{\dot{\hat{X}}}} = A_1 \underset{X}{\underset{\downarrow}{\hat{X}}} + \phi_1(\underset{X}{\underset{\downarrow}{\hat{X}}}) + B \underset{u}{\underset{\downarrow}{\hat{u}}} + \Gamma(y - \hat{y}). \tag{5.87}$$

The controller u can guarantee that the state errors e_p, e_5, and e_6 are asymptotic convergence to 0 after the Lyapunov candidate functions V_3, V_4, and V_5 are constructed. Moreover, the state errors e_1 and e_2 are also asymptotic convergence to 0 on the basis of Equation 5.81. Obviously, the matrix A_e in Equation 5.84 is Hurwitz. Therefore, the full-state error dynamics shown in Equation 5.83 is a quasi-steady-state model as follows:

$$\dot{Z} = A_e Z + B_e Z + B(u - \hat{u}) \overset{t \geq \varsigma}{\Leftrightarrow} \dot{Z} = A_e Z + B_e Z, \tag{5.88}$$

where ς is a definite time constant of t.

When $t \to \infty$, according to Equation 5.88, the full-state error Z converges to 0 as quickly as possible.

5.4 Experiment

To test the proposed backstepping control method and the full-state observer associated with the control of EHS, the experimental framework of the model is presented as shown in Figure 3.16. Some of the same hydraulic parameters are shown in Table 3.9. The other parameters are $V_{0a} = 1.74 \times 10^{-5}\,\mathrm{m}^3$, $V_{0b} = 8.66 \times 10^{-6}\,\mathrm{m}^3$, $K_{sv} = 7.9 \times 10^{-4}\,\mathrm{m/V}$, $\omega_{sv} = 353.6\,\mathrm{rad/s}$, $\zeta_{sv} = 0.707$, $b = 2500\,\mathrm{Ns/m}$, $K = 1000\,\mathrm{N/m}$, $F_{fstatic} = 20\,\mathrm{N}$, $\Delta_{F_L} = 500\,\mathrm{N}$, and $p_b = 20\%$. Since the stroke of the cylinder is 79 mm and the shoulder angle of the upper arm is from $-70°$ to $50°$, the range of absolute coordinates position x_1 is from -0.0476 to $0.0303\,\mathrm{m}$. However, in the experiment, the displacement of the cylinder is not more than 58 mm to avoid the boundary collision of the robotic arm.

The observer and control parameters are designed as follows:

1. The observer parameters: $\lambda = 0.5$, $\Gamma_1 = 1000$, $\Gamma_2 = 10$, $\Gamma_3 = 10^{-3}$, $\Gamma_4 = 10^{-3}$, $\Gamma_5 = 1000$, $\Gamma_6 = 10$.
2. The control parameters: $k_1 = 10^{-6}$, $k_2 = 10^{-9}$, $k_3 = 10^{-12}$, $\sigma_1 = 500$.
3. In the function tanh(.), $k = 1000$.

5.4.1 Result of the Proposed Method

The full states are first estimated by observer Equation 5.25. Then, Equations 5.54, 5.55, and 5.71 or 5.73, and the estimated states are substituted into the controller (5.78) to compute the current control voltage. Here, the displacement command is considered as two types: sinusoidal and step demands. The amplitude of the sinusoidal command is 0.0289 m with frequency 0.5 Hz. The step command also has the same amplitude. The actual displacement can be obtained by trigonometry of the robotic arm according to the angle measured by the relative encoder.

Figures 5.1 through 5.4 show the related results of the sinusoidal and step experiment, respectively. In Figure 5.1a, after 1.27 s, the actual cylinder displacement has been close to the value 0.025 m when the steady error is less than 5%. The maximum control voltage is 9 V, not more than its saturation ±10 V as shown in Figure 5.1b. When the displacement nears the step command, the control voltage reduces to 0 quickly. The estimation of x_1 is a correct prediction for the actual cylinder displacement by the designed observer as shown in Figure 5.2a. According to the estimations of x_3 and x_4 as shown in Figure 5.2c and d, the dynamic estimation is sharp due to the existing error of the high observer. Moreover, this single-rod EHS has a one-dimensional internal dynamics, which results in the rigid dynamic error of estimated state. This phenomenon is also shown in the sinusoidal experiment, which means the estimation of load pressure p_L defined in Equation 5.51 is not smooth until the dynamic response finishes. In step experiment, after approximately 3 s later, the estimation of p_L is smooth and is close to 30 bar shown in Figure 5.2c. The estimation of spool position x_5 is corresponded to the control voltage $\hat{u}$, which is not more than its saturation ±7.9 mm shown in Figure 5.2e. In the sinusoidal experiment, the maximum dynamic error of displacement is less than 5.7 mm, which means the dynamic error is not more than

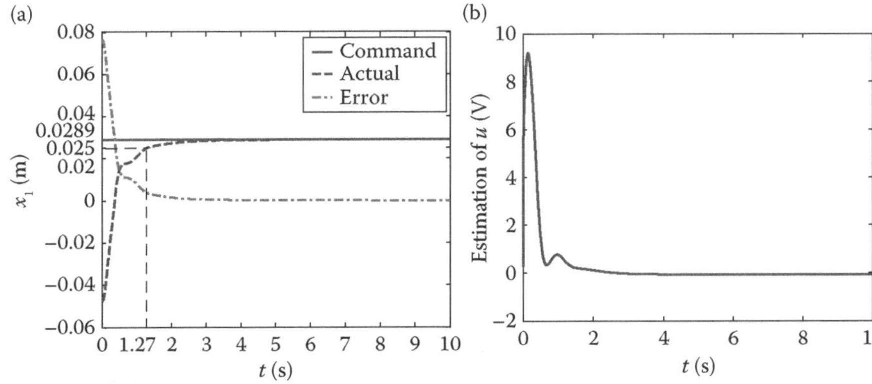

FIGURE 5.1
Step response experiment of the upper arm hydraulic actuator. (a) The position of the upper arm cylinder. (b) The control voltage of the upper arm cylinder.

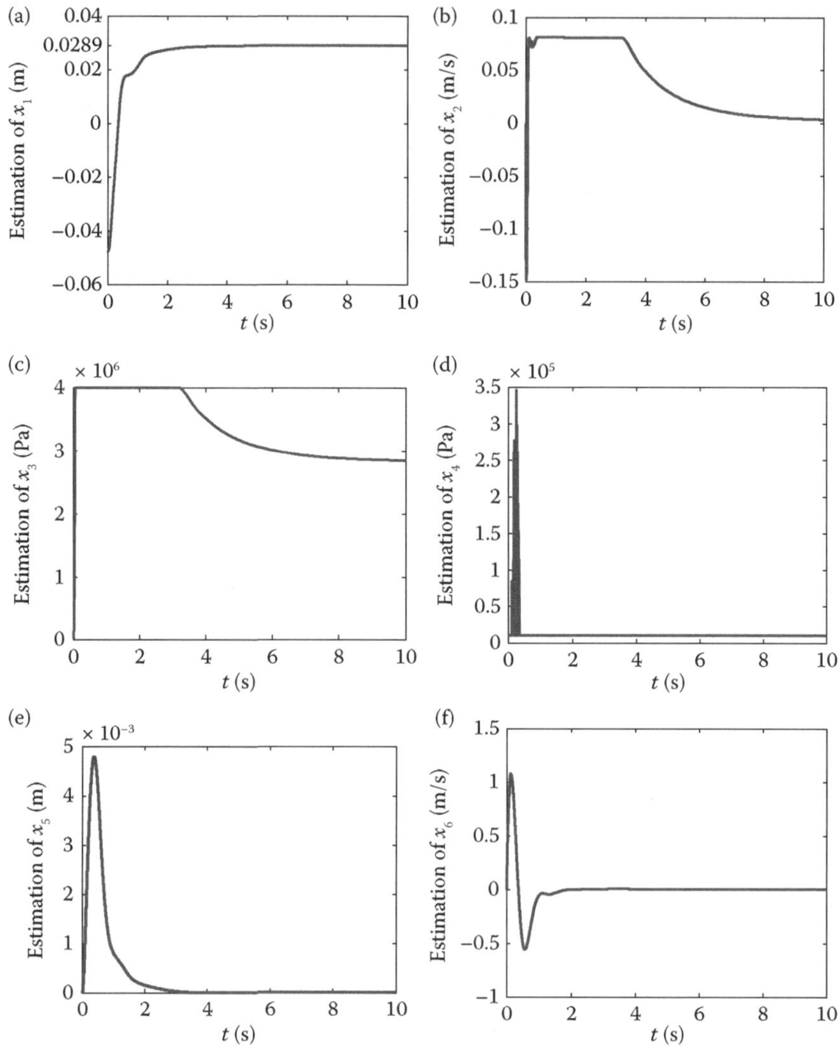

FIGURE 5.2
State estimations by the high-gain state observer in step response experiment. (a) Estimation of x_1 in step response. (b) Estimation of x_2 in step response. (c) Estimation of x_3 in step response. (d) Estimation of x_4 in step response. (e) Estimation of x_5 in step response. (f) Estimation of x_6 in step response.

10% when the sinusoidal frequency is 0.5 Hz. Owing to the initial position error of x_1 and dynamic estimated error for the sinusoidal command, the control voltage is saturation in a short duration. When the actual displacement tracks the command well, the control voltage is less than its saturation.

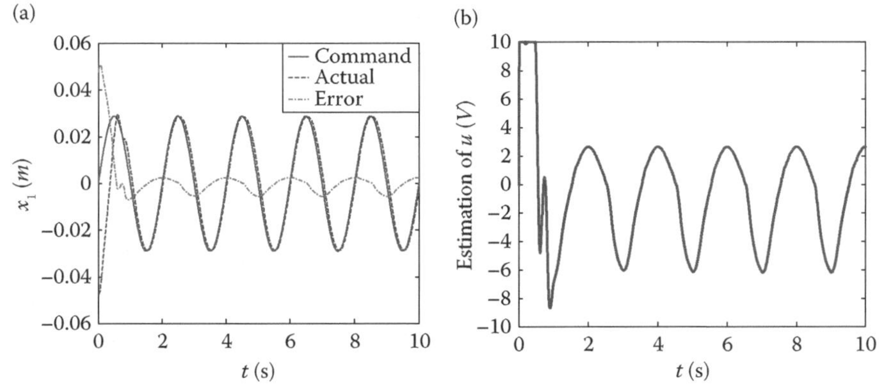

FIGURE 5.3
Sinusoidal response experiment of the upper arm hydraulic actuator. (a) The position of the upper arm cylinder. (b) The control voltage of the upper arm cylinder.

5.4.2 Compared Result

The conventional PI control method $u = k_p(x_{1d} - x_1) + k_i \int (x_{1d} - x_1)dt$ is also used in this experiment. To illustrate the problem, we compare the PI method and the backstepping method in two different conditions:

1. The sinusoidal frequency is 0.5 Hz and the elbow is always retracted as seen in Figure 5.8a, which means the required control bandwidth is low and the dynamic external load is relatively small.
2. The sinusoidal frequency is 1 Hz and the elbow is always extended as seen in Figure 5.8b, which is the worst control condition for the shoulder hydraulic actuator.

Here, we choose the PI control parameters $k_p = 100$ and $k_i = 15$, which can guarantee the fast response of displacement shown in Figure 5.5a. In fact, the dynamic error of the PI method is smaller than the proposed method in the first condition shown in Figure 5.5b. But when the required frequency of command is increased and the external load nears the critical value, which is the maximum bearing load of the shoulder hydraulic actuator, the dynamic performance of the PI method becomes worse than the proposed method shown in Figure 5.6a. The control voltage is more intense for the PI method as shown in Figure 5.6c, which results into the dynamic displacement error that is larger than the proposed method shown in Figure 5.6b. These compared results show that the fixed control parameters of the PI method cannot fit all work conditions. If the control parameters are chosen big enough, the robustness performance will become weak in critical conditions, except most normal conditions. Therefore, the proposed method is suitable for this robotic

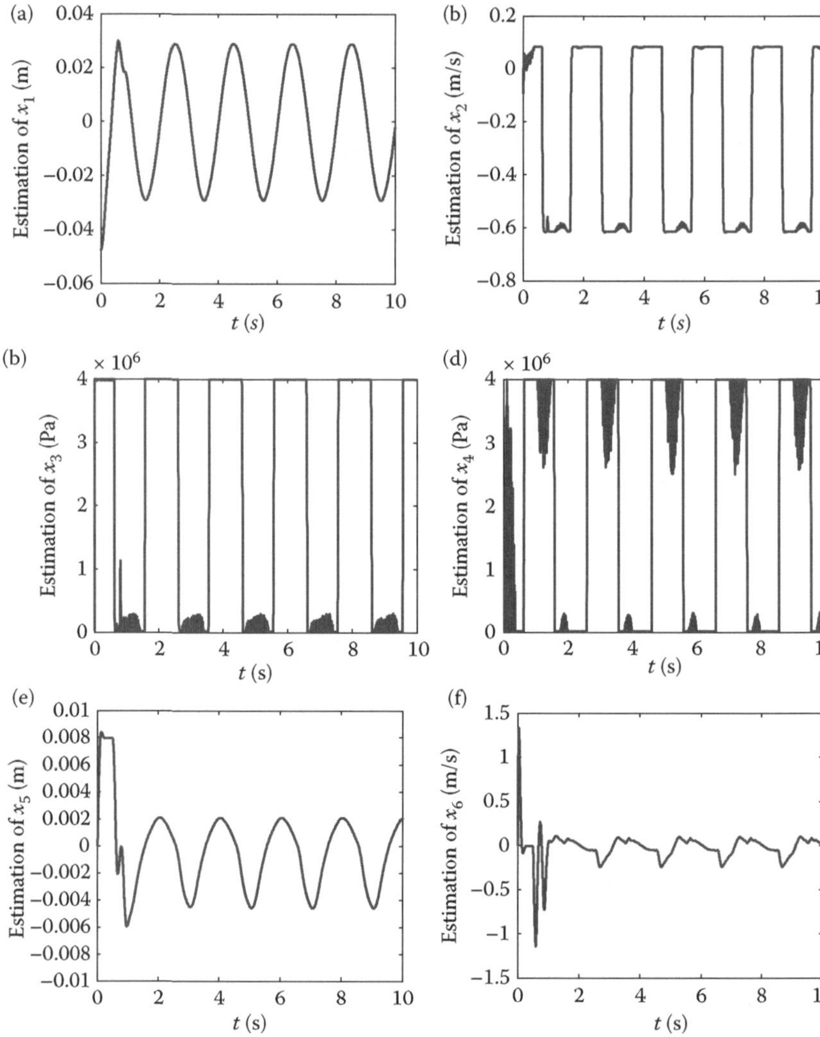

FIGURE 5.4
State estimations by the high-gain state observer in the sinusoidal response experiment. (a) Estimation of x_1 in the sinusoidal response. (b) Estimation of x_2 in the sinusoidal response. (c) Estimation of x_3 in the sinusoidal response. (d) Estimation of x_4 in the sinusoidal response. (e) Estimation of x_5 in the sinusoidal response. (f) Estimation of x_6 in the sinusoidal response.

arm in some critical conditions where the advantage of the PI method is limited.

Here, two joints of this robotic arm can be driven simultaneously to realize the coordinated motion. The hydraulic actuator of the shoulder is controlled by the proposed method and the actuator of the elbow is controlled by the

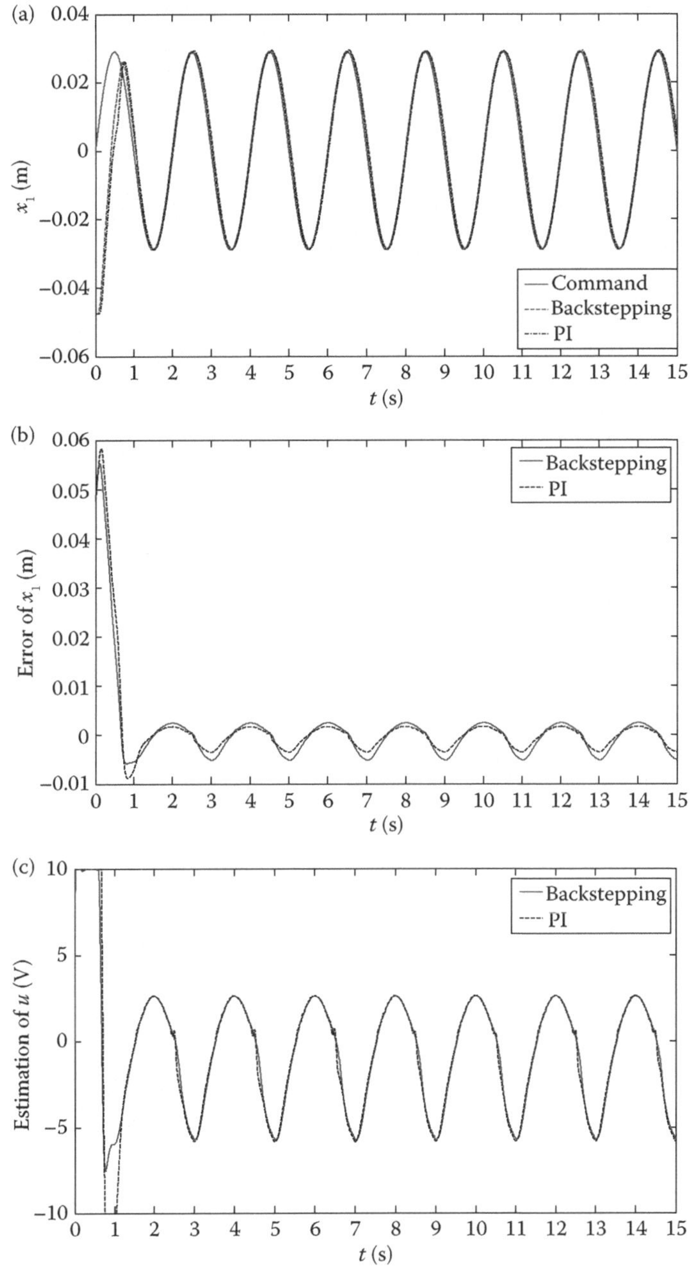

FIGURE 5.5
Comparison result in condition (1). (a) The cylinder position response of the sinusoidal demand. (b) The cylinder position error of the sinusoidal demand. (c) The control voltage estimation of the sinusoidal demand.

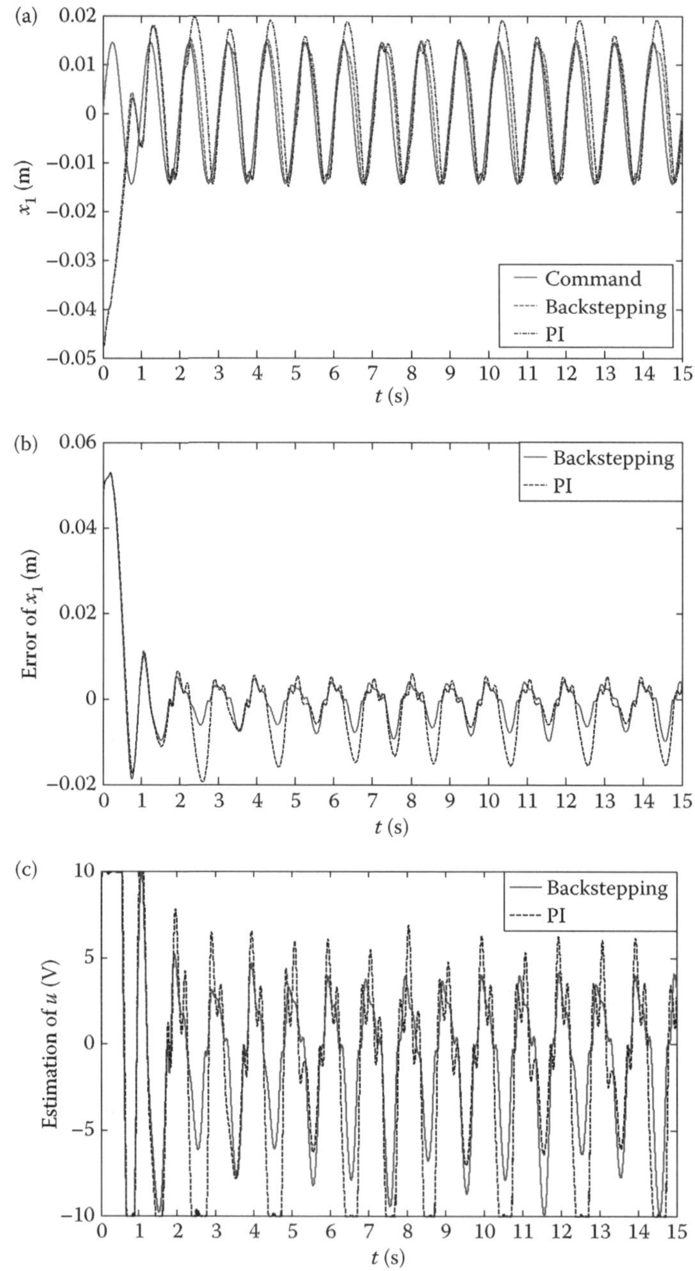

FIGURE 5.6
Comparison result in condition (2). (a) The cylinder position response of the sinusoidal demand. (b) The cylinder position error of the sinusoidal demand. (c) The control voltage estimation of the sinusoidal demand.

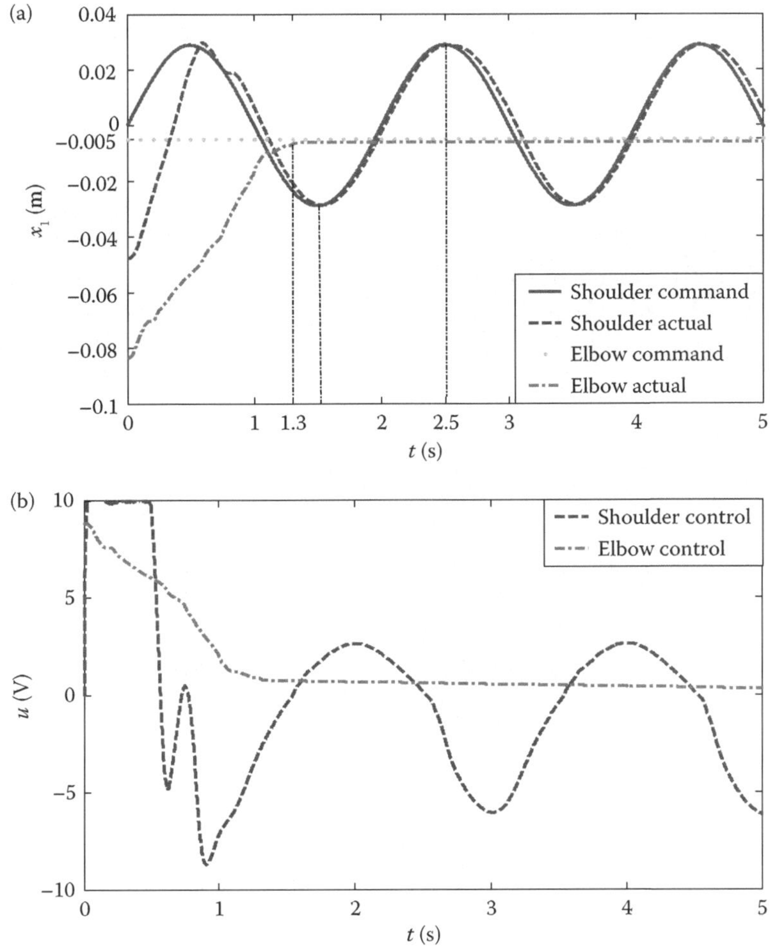

FIGURE 5.7
The coordinated motion experiment results of the robotic arm joints, sinusoidal demand input for the shoulder actuator, and step demand input for the elbow actuator. (a) The two cylinder positions response of sinusoidal and step demands. (b) The two control voltages of sinusoidal and step demands.

PI method. The two input displacement commands are sinusoidal and step, respectively, which is the same command as section B. The displacement response and the corresponding control voltage of two hydraulic actuators are shown in Figure 5.7. At initial time 0 s, the two joints are all retracted. Then, after 1.3 s, the joint angle of the elbow is extended to its maximum 130° by the PI control and then the elbow joint will always keep this extended state. After 1.5 s, the shoulder actuator is retracted to its minimum displacement − 0.0289 m. After 2.5 s, the shoulder actuator is extended to its

(a) (b) (c) (d)

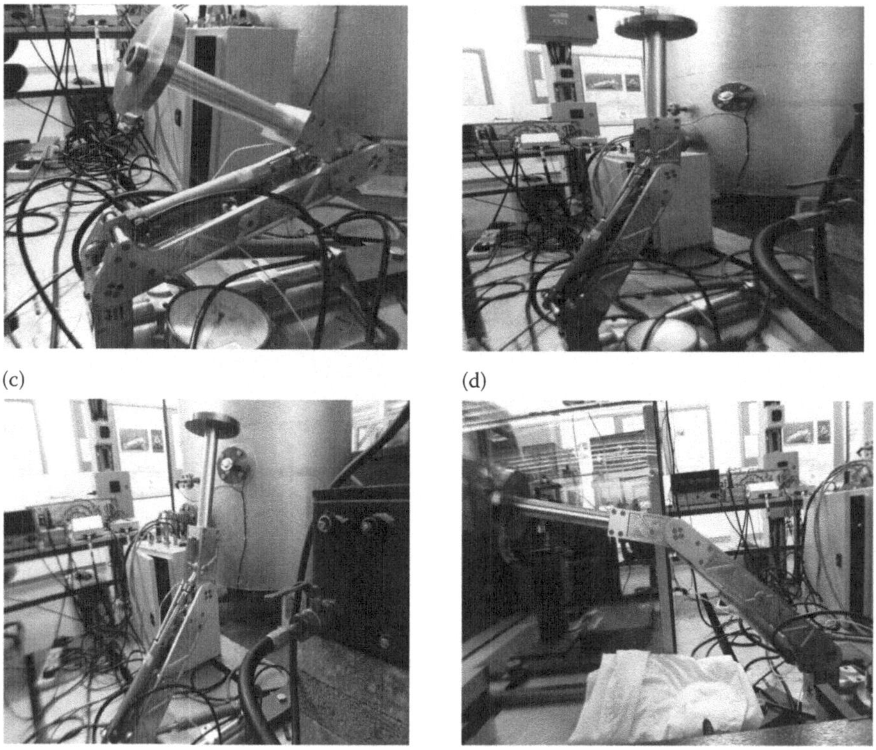

FIGURE 5.8
Experiment video of coordinated motion for the robotic arm joints. (a) $t = 0\,\text{s}$. (b) $t = 1.3\,\text{s}$. (c) $t = 1.5\,\text{s}$. (d) $t = 2.5\,\text{s}$.

maximum displacement 0.0289 m. Then the sinusoidal law of the shoulder joint is repeated by the proposed control method. The experiment video of coordinated motion for the robotic arm joints is shown in Figure 5.8, which is corresponding to the four sample points.

6

Parametric Adaptive Control Method

CONTENTS

To the best of the authors' knowledge, the aforementioned backstepping and adaptive controllers need to deal with some derivatives of virtual control variables, which exist in backstepping iteration. These derivatives can be computed by the system state error model and parametric estimated law established in recursive controller design [37]. Since this computed process easily leads to derivative explosion, few research are focused on the quality of these virtual control variables (i.e., data validation). If these virtual control variables are not available, especially violent, both the stability and dynamic performances of EHS will be significantly decreased. Thus, in this chapter, a parametric adaptive backstepping control method [18] is presented based on state feedback to estimate some unknown parameters in a hydraulic model. Then, a decayed memory filter is proposed to compute the derivatives value of virtual control variables in the backstepping control design. The effectiveness of the proposed control is verified by a comparative experimental study.

6.1 Dynamic Model of EHS

For convenient illustration, the dynamic model of EHS is adopted by the symmetrical hydraulic actuator. The load flow Q_L of the servo valve mentioned in Equation 2.5 is given by

$$Q_L = C_d w x_v \sqrt{\frac{1}{\rho}\left(p_s - \mathrm{sgn}\,(x_v)\, p_L\right)}, \tag{6.1}$$

where p_L is the load pressure of the cylinder, x_v is the spool position of the servo valve, p_s is the supply pressure of the pump, C_d is the discharge coefficient, w is the area gradient of the valve spool, and ρ is the density of the hydraulic oil.

The flow-pressure continuous equation of the hydraulic cylinder mentioned in Equation 2.1 is given by

$$Q_L = A_p \frac{dy}{dt} + C_{tl} p_L + \frac{V_t}{4\beta_e}\frac{dp_L}{dt}, \tag{6.2}$$

where y is the displacement of the cylinder, C_{tl} is the coefficient of the total leakage of the cylinder, β_e is the effective bulk modulus, A_p is the annulus area of the cylinder chamber, and V_t is the half-volume of the cylinder.

If the major viscous friction of the hydraulic oil is simplified as Coulomb friction, the mechanical dynamic equation driven by the hydraulic actuator mentioned in Equation 6.3 is shown as follows:

$$A_p p_L = m\ddot{y} + b\dot{y} + Ky + F_L, \tag{6.3}$$

where m is the load mass, K is the load spring constant, b is the viscous damping coefficient, and F_L is the external load on the hydraulic actuator.

The dynamics of the servo valve is adopted by one-order linear model mentioned in Equation 2.8, which is given by

$$T_{sv}\dot{x}_v + x_v = K_{sv}u, \tag{6.4}$$

where K_{sv} is the gain of the control voltage u and T_{sv} is the response time constant of the servo valve.

Remark 6.1

In Equation 6.1, the function sgn(.) should be smoothed in the derivation of backstepping control, which is replaced by the hyperbolic tangent function tanh(.) as follows [35]:

$$\mathrm{sgn}(x_v) \approx \tanh(kx_v) = \frac{e^{kx_v} - e^{-kx_v}}{e^{kx_v} + e^{-kx_v}}, \quad k \gg 0, \tag{6.5}$$

where k is a positive constant.

If the state vector $X = [x_1, x_2, x_3, x_4] = [y, \dot{y}, p_L, x_v]^T$, the output variable $y = x_1$, and the control input $u(t)$ is defined, the dynamics of EHS is constructed as the fourth state space model as follows:

$$
\begin{cases}
\dot{x}_1 = x_2 \\
\dot{x}_2 = \dfrac{1}{m}(-Kx_1 - bx_2 + A_p x_3 - F_L) \\
\dot{x}_3 = -\dfrac{4\beta_e A_p}{V_t} x_2 - \dfrac{4\beta_e C_{tl}}{V_t} x_3 + \dfrac{4\beta_e C_d w}{V_t \sqrt{\rho}} \sqrt{p_s - \tanh(kx_4) x_3 x_4} \\
\dot{x}_4 = -\dfrac{1}{T_{sv}} x_4 + \dfrac{K_{sv}}{T_{sv}} u
\end{cases}
\quad . \tag{6.6}
$$

The external load $F_L(t)$ is divided into two elements, which is mentioned in Equation 4.11 as follows:

$$
\begin{cases}
F_{Lu}(\theta_1, \theta_2) = \dfrac{T_u(\theta_1, \theta_2)}{l_1(\theta_1)} \\
F_{Lf}(\theta_1, \theta_2) = \dfrac{T_f(\theta_1, \theta_2)}{l_2(\theta_2)}
\end{cases}
, \tag{6.7}
$$

where F_{Lu} is the load force on the shoulder hydraulic actuator, F_{Lf} is the load force on the elbow hydraulic actuator, and the dynamic force arms $l_i(\theta_i)$, $(i = 1, 2)$ are computed by triangle geometry mentioned in Equation 3.48.

Remark 6.2

Owing to measurement noise and uncertainty disturbance in engineering practice, F_{Lu}, F_{Lf} are difficult to be obtained. The computed value in Equation 6.7 may deviate from the true value. Thus, these values should be handled by a filter before being used in parametric estimation law.

6.2 Parametric Adaptive Backstepping Controller

All the states $x_i (i = 1, \ldots, 4)$ are known for the controller design. If these unknown parameters are defined as $\vartheta_1 = K$, $\vartheta_2 = b$, $\vartheta_3 = F_L$, $\vartheta_4 = ((4\beta_e A_p)/V_t)$, $\vartheta_5 = ((4\beta_e C_{tl})/V_t)$, $\vartheta_6 = ((4\beta_e C_d w)/(V_t \sqrt{\rho}))$, then the dynamics of EHS can be described as follows [35,36]:

$$\begin{cases} \dot{x}_1 = x_2 \\ \dot{x}_2 = \dfrac{A_p}{m}x_3 - \dfrac{\vartheta_1}{m}x_1 - \dfrac{\vartheta_2}{m}x_2 - \dfrac{\vartheta_3}{m} \\ \dot{x}_3 = -\vartheta_4 x_2 - \vartheta_5 x_3 + \vartheta_6\sqrt{p_s - \tanh(kx_4)x_3x_4} \\ \dot{x}_4 = -\dfrac{1}{T_{sv}}x_4 + \dfrac{K_{sv}}{T_{sv}}u \end{cases}. \tag{6.8}$$

6.2.1 General Backstepping Control Design

At first, the state error $z_i(i = 1,\ldots,4)$ and parametric estimated error $\tilde{\vartheta}_i(i = 1,\ldots,6)$ are defined as follows:

$$\begin{cases} z_1 = x_1 - x_{1d} \\ z_i = x_i - \alpha_{i-1}(i = 2,3,4) \\ \tilde{\vartheta}_i = \vartheta_i - \hat{\vartheta}_i, (i = 1,\ldots,6) \end{cases}, \tag{6.9}$$

where x_{1d} is the demand of cylinder displacement, α_i is the ith virtual control, and $\hat{\vartheta}_i$ is the estimation of ϑ_i.

Theorem 6.1

Assuming that the unknown parameters $\vartheta_i(i = 1,\ldots,6)$ are constant, if the virtual control variables $\alpha_i \in C^1(i = 1,2,3)$ and their derivatives $\dot{\alpha}_i(i = 1,2,3)$ are smooth, then there exists a backstepping controller u, which guarantees $z_i(t)(i = 1,\ldots,4) \to 0$, $\tilde{\vartheta}_i(t)(i = 1,\ldots,6) \to 0$, as $t \to \infty$.

Proof. The derivative of z_1 is given by

$$\dot{z}_1 = z_2 + \alpha_1 - \dot{x}_{1d}. \tag{6.10}$$

Then the Lyapunov function V_1 is defined as

$$V_1 = \frac{1}{2}z_1^2. \tag{6.11}$$

Since the derivative of V_1 is

$$\dot{V}_1 = z_1(z_2 + \alpha_1 - \dot{x}_{1d}), \tag{6.12}$$

the virtual control α_1 is designed as follows:

$$\alpha_1 = -c_1 z_1 + \dot{x}_{1d}, \tag{6.13}$$

where c_1 is a positive constant.

Therefore, the derivative of V_1 is

$$\dot{V}_1 = -c_1 z_1^2 + z_1 z_2. \tag{6.14}$$

Since the derivative of z_2 is

$$\begin{aligned}
\dot{z}_2 &= \dot{x}_2 - \dot{\alpha}_1 \\
&= \frac{A_p}{m} x_3 - \frac{\vartheta_1}{m} x_1 - \frac{\vartheta_2}{m} x_2 - \frac{\vartheta_3}{m} + c_1 x_2 - c_1 \dot{x}_{1d} - \ddot{x}_{1d} \\
&= \frac{A_p}{m} z_3 + \frac{A_p}{m} \alpha_2 - \frac{\vartheta_1}{m} x_1 - \frac{\vartheta_3}{m} x_2 - \frac{\vartheta_3}{m} + c_1 x_2 - c_1 \dot{x}_{1d} - \ddot{x}_{1d},
\end{aligned} \tag{6.15}$$

the Lyapunov function V_2 is defined as follows:

$$V_2 = V_1 + \frac{1}{2} z_2^2 + \sum_{i=1}^{3} \frac{1}{2k_i} \left(\vartheta_i - \hat{\vartheta}_i \right)^2, \tag{6.16}$$

where k_1, k_2, and k_3 are positive constants.

Therefore, the derivative of V_2 is

$$\begin{aligned}
\dot{V}_2 &= -c_1 z_1^2 + z_1 z_2 + z_2 \left(\frac{A_p}{m} z_3 + \frac{A_p}{m} \alpha_2 - \frac{\vartheta_1}{m} x_1 - \frac{\vartheta_2}{m} x_2 - \frac{\vartheta_3}{m} \right. \\
&\quad \left. + c_1 x_2 - c_1 \dot{x}_{1d} - \ddot{x}_{1d} \right) - \frac{\dot{\hat{\vartheta}}_1}{k_1} \tilde{\vartheta}_1 - \frac{\dot{\hat{\vartheta}}_2}{k_2} \tilde{\vartheta}_2 - \frac{\dot{\hat{\vartheta}}_3}{k_3} \tilde{\vartheta}_3 \\
&= -c_1 z_1^2 + z_2 \left(z_1 + \frac{A_p}{m} z_3 + \frac{A_p}{m} \alpha_2 - \hat{\vartheta}_1 x_1 / m - \hat{\vartheta}_2 x_2 / m - \hat{\vartheta}_3 / m \right. \\
&\quad \left. + c_1 x_2 - c_1 \dot{x}_{1d} - \ddot{x}_{1d} \right) - \tilde{\vartheta}_1 \left(\frac{\dot{\hat{\vartheta}}_1}{k_1} + x_1 z_2 / m \right) - \tilde{\vartheta}_2 \left(\frac{\dot{\hat{\vartheta}}_2}{k_2} + x_2 z_2 / m \right) \\
&\quad - \tilde{\vartheta}_3 \left(\frac{\dot{\hat{\vartheta}}_3}{k_3} + z_2 / m \right).
\end{aligned} \tag{6.17}$$

If the parametric adaptive estimation laws and the virtual control α_2 are designed as follows:

$$\dot{\hat{\vartheta}}_1 = -k_1 x_1 z_2 / m, \quad \dot{\hat{\vartheta}}_2 = -k_2 x_2 z_2 / m, \quad \dot{\hat{\vartheta}}_3 = -k_3 z_2 / m, \tag{6.18}$$

$$\alpha_2 = -\frac{m}{A_p} z_1 - \frac{c_2 m}{A_p} z_2 + \frac{m}{A_p} \left(\hat{\vartheta}_1 x_1 / m + \hat{\vartheta}_2 x_2 / m + \hat{\vartheta}_3 / m - c_1 x_2 + c_1 \dot{x}_{1d} + \ddot{x}_{1d} \right), \tag{6.19}$$

we see that the derivative of V_2 is

$$\dot{V}_2 = -c_1 z_1^2 - c_2 z_2^2 + \frac{A_p}{m} z_2 z_3, \tag{6.20}$$

where c_2 is a positive constant.

Then the derivative of z_3 is

$$\dot{z}_3 = \dot{x}_3 - \dot{\alpha}_2 = -\vartheta_4 x_2 - \vartheta_5 x_3 + \vartheta_6 \sqrt{p_s - \tanh(kx_4) x_3}\, x_4 - \dot{\alpha}_2. \tag{6.21}$$

If the Lyapunov function V_3 is defined as

$$V_3 = V_2 + \frac{1}{2} z_3^2 + \sum_{i=4}^{6} \frac{1}{2k_i} \left(\vartheta_i - \hat{\vartheta}_i\right)^2, \tag{6.22}$$

then the derivative of V_3 is

$$\begin{aligned}
\dot{V}_3 &= -c_1 z_1^2 - c_2 z_2^2 + \frac{A_p}{m} z_2 z_3 \\
&\quad + z_3 \left(-\vartheta_4 x_2 - \vartheta_5 x_3 + \vartheta_6 \sqrt{p_s - \tanh(kx_4) x_3}\,(z_4 + \alpha_3) - \dot{\alpha}_2 \right) \\
&\quad - \frac{\dot{\hat{\vartheta}}_4}{k_4} \tilde{\vartheta}_4 - \frac{\dot{\hat{\vartheta}}_5}{k_5} \tilde{\vartheta}_5 - \frac{\dot{\hat{\vartheta}}_6}{k_6} \tilde{\vartheta}_6 \\
&= -c_1 z_1^2 - c_2 z_2^2 + z_3 \left(\begin{array}{l} \frac{A_p}{m} z_2 + \hat{\vartheta}_6 \sqrt{p_s - \tanh(kx_4) x_3}\, z_4 \\ + \hat{\vartheta}_6 \sqrt{p_s - \tanh(kx_4) x_3}\, \alpha_3 - \hat{\vartheta}_4 x_2 - \hat{\vartheta}_5 x_3 - \dot{\alpha}_2 \end{array} \right) \\
&\quad - \tilde{\vartheta}_4 \left(\frac{\dot{\hat{\vartheta}}_4}{k_4} + x_2 z_3 \right) - \tilde{\vartheta}_5 \left(\frac{\dot{\hat{\vartheta}}_5}{k_5} + x_3 z_3 \right) - \tilde{\vartheta}_6 \left(\frac{\dot{\hat{\vartheta}}_6}{k_6} - x_4 z_3 \sqrt{p_s - \tanh(kx_4) x_3} \right),
\end{aligned} \tag{6.23}$$

where k_4, k_5, and k_6 are positive constants.

If the parametric adaptive estimation laws and the virtual control α_3 are designed as follows:

$$\dot{\hat{\vartheta}}_4 = -k_4 x_2 z_3, \quad \dot{\hat{\vartheta}}_5 = -k_5 x_3 z_3, \quad \dot{\hat{\vartheta}}_6 = k_6 x_4 z_3 \sqrt{p_s - \tanh(kx_4) x_3}, \tag{6.24}$$

$$\alpha_3 = -\frac{\frac{m}{A_p} z_2 + c_3 z_3 + \hat{\vartheta}_4 x_2 + \hat{\vartheta}_5 x_3 + \dot{\alpha}_2}{\hat{\vartheta}_6 \sqrt{p_s - \tanh(kx_4) x_3}}, \tag{6.25}$$

we see that the derivative of V_3 is

$$\dot{V}_3 = -c_1 z_1^2 - c_2 z_2^2 - c_3 z_3^2 + z_3 z_4 \hat{\vartheta}_6 \sqrt{p_s - \tanh(kx_4) x_3}, \tag{6.26}$$

where c_3 is a positive constant.

If the Lyapunov function V_4 is defined as

$$V_4 = V_3 + \frac{1}{2}z_4^2, \tag{6.27}$$

then the derivative of V_4 is

$$\begin{aligned}\dot{V}_4 &= \dot{V}_3 + z_4\dot{z}_4 = -c_1z_1^2 - c_2z_2^2 - c_3z_3^2 \\ &\quad + z_3z_4\hat{\vartheta}_6\sqrt{p_s - \tanh(kx_4)x_3} + z_4(\dot{x}_4 - \dot{\alpha}_3) \\ &= -c_1z_1^2 - c_2z_2^2 - c_3z_3^2 + z_4\left(z_3\hat{\vartheta}_6\sqrt{p_s - \tanh(kx_4)x_3} - \frac{x_4}{T_{sv}} + \frac{K_{sv}}{T_{sv}}u - \dot{\alpha}_3\right).\end{aligned} \tag{6.28}$$

If the controller is designed as

$$u = -\frac{T_{sv}}{K_{sv}}c_4z_4 + \frac{x_4}{K_{sv}} + \frac{T_{sv}}{K_{sv}}\dot{\alpha}_3 - \frac{T_{sv}}{K_{sv}}z_3\hat{\vartheta}_6\sqrt{p_s - \tanh(kx_4)x_3}, \tag{6.29}$$

then the derivative of V_4 is

$$\dot{V}_4 = -c_1z_1^2 - c_2z_2^2 - c_3z_3^2 - c_4z_4^2 < 0, \tag{6.30}$$

where c_4 is a positive constant.

Thus, by the parametric adaptive estimation laws (6.18) and (6.24), the elements involving $\tilde{\vartheta}_i(i = 1, \ldots, 6)$ vanish in $\dot{V}_4$. Simultaneously, $\dot{V}_4$ is negative definite by the virtual control (6.13), (6.19), and (6.25) and the backstepping controller (6.29). This iteration controller guarantees the parametric estimation errors $\tilde{\vartheta}_i(i = 1, \ldots, 6)$ and the system state error $z_i(i = 1, \ldots, 4)$ converging to zero. ■

From the assumption in Theorem 6.2, the derivative of α_2, α_3 used in Equations 6.19 and 6.25 can be simplified as follows:

$$\begin{aligned}\dot{\alpha}_2 &= -\frac{m}{A_p}(z_2 + \alpha_1 - \dot{x}_{1d}) - c_2z_3 - c_2\alpha_2 \\ &\quad + \frac{c_2m}{A_p}(-\hat{\vartheta}_1x_1/m - \hat{\vartheta}_2x_2/m - \hat{\vartheta}_3/m + c_1x_2 - c_1\dot{x}_{1d} - \ddot{x}_{1d}) - \frac{c_1m}{A_p}\hat{\dot{x}}_2,\end{aligned} \tag{6.31}$$

$$\dot{\alpha}_3 = -\frac{z_3 - \frac{m}{A_p}(c_2z_2 + z_1) + c_3(\hat{\dot{x}}_3 - \dot{\alpha}_2) + \hat{\vartheta}_4\hat{\dot{x}}_2 + \hat{\vartheta}_5\hat{\dot{x}}_3 + \ddot{\alpha}_2}{\hat{\vartheta}_6\sqrt{p_s - \tanh(kx_4)x_3}}, \tag{6.32}$$

where $\hat{x}_2, \hat{x}_3$ is estimated by the dynamic model of EHS as follows:

$$\begin{cases} \hat{\dot{x}}_2 = \dfrac{A_p}{m} x_3 - \hat{\vartheta}_1 x_1/m - \hat{\vartheta}_2 x_2/m - \hat{\vartheta}_3/m \\ \hat{\dot{x}}_3 = -\hat{\vartheta}_4 x_2 - \hat{\vartheta}_5 x_3 + \hat{\vartheta}_6 \sqrt{p_s - \tanh(kx_4) x_3 x_4} \end{cases} . \quad (6.33)$$

6.2.2 Decayed Memory Filter Design

As shown in Equations 6.31 and 6.32, although the derivatives $\dot{\alpha}_2, \dot{\alpha}_3$ are computed in backstepping design, the algorithm is still very complicated. If these derivatives are directly used in the controller u, the robustness of this backstepping controller may decline. Therefore, a decayed memory filter is designed to re-estimate these derivatives and filter the variable parameter ϑ_3 mentioned in Remark 6.1. The filter form is described as follows [65]:

$$\begin{cases} E_\phi(k) = \phi(k) - \left[\hat{\phi}(k-1) + \hat{\dot{\phi}}(k-1) \cdot T_c\right] \\ \hat{\phi}(k) = \hat{\phi}(k-1) + \hat{\dot{\phi}}(k-1) \cdot T_c + G \cdot E_\phi(k) \\ \hat{\dot{\phi}}(k) = \hat{\dot{\phi}}(k-1) + \dfrac{H}{T_c} E_\phi(k) \end{cases} , \quad \phi = \alpha_2, \alpha_3, \vartheta_3 , \quad (6.34)$$

where $\phi(k)$ is the kth calculated value of ϕ, $\hat{\phi}(k)$ is the estimation of $\phi(k)$, $\hat{\dot{\phi}}(k)$ is the kth estimation of the derivative $\dot{\phi}$, T_c is the interval of the controller, the filter parameters $G = 1 - \lambda^2$, $H = (1 - \lambda)^2$, and the range of the filter factor λ is (0, 1).

Since the calculated values of α_2, α_3 are obtained by Equations 6.19 and 6.25, the estimation values $\hat{\alpha}_i, \hat{\dot{\alpha}}_i (i = 2, 3)$ are obtained by Equation 6.34.

From Equation 6.7, the calculated value of ϑ_3 is described as follows:

$$\begin{cases} \vartheta_{31} = \dfrac{F_{Lu}(\theta_1, \theta_2)}{m_{1f}} = \dfrac{T_u(\theta_1, \theta_2)}{m_{1f} l_1(\theta_1)} \\ \vartheta_{32} = \dfrac{F_{Lf}(\theta_1, \theta_2)}{m_{2f}} = \dfrac{T_f(\theta_1, \theta_2)}{m_{2f} l_2(\theta_2)} \end{cases} , \quad (6.35)$$

where ϑ_{31}, ϑ_{32} represent two calculated values of the parameter ϑ_3 for two hydraulic actuators. After Equation 6.35 is substituted into Equation 6.34, the estimation values $\hat{\vartheta}_{3i}, \hat{\dot{\vartheta}}_{3i} (i = 1, 2)$ are also obtained.

Then the convergence of the proposed filter should be analyzed. If the vector $\chi(k)$ is defined as $\chi(k) = \left[\hat{\alpha}(k), \hat{\dot{\alpha}}(k)\right]^T$, then the filter (6.34) is described as follows:

$$\begin{bmatrix} \hat{\alpha}(k) \\ \hat{\dot{\alpha}}(k) \end{bmatrix} = \begin{bmatrix} 1-G & (1-G)T_c \\ -\dfrac{H}{T_c} & 1-H \end{bmatrix} \begin{bmatrix} \hat{\alpha}(k-1) \\ \hat{\dot{\alpha}}(k-1) \end{bmatrix} + \begin{bmatrix} G \\ \dfrac{H}{T_c} \end{bmatrix} \alpha(k) . \tag{6.36}$$

Theorem 6.2

The input $\alpha(k)$ is assumed to be the input of the discrete linear system (6.36), and $\hat{\alpha}(k)$, $\hat{\dot{\alpha}}(k)$ are the outputs; then Equation 6.36 is input-to-state stable (ISS), and $\hat{\alpha}(k) \to \alpha(k)$, $\hat{\dot{\alpha}}(k) \to \dot{\alpha}(k)$, as $k \to \infty$.

Proof.

1. If two constants are defined as $\lambda_1 = 1 - G$, $\lambda_2 = 1 - H$, then we can see that $\lambda_1 < 1$, $\lambda_2 < 1$, $\lambda_1 < \lambda_2$. The characteristic polynomial of system matrix A in Equation 6.36 is described as follows:

$$|sI - A| = \begin{bmatrix} s - \lambda_1 & -\lambda_1 T_c \\ -\dfrac{\lambda_2 - 1}{T_c} & s - \lambda_2 \end{bmatrix} = s^2 - (\lambda_1 + \lambda_2)s + \lambda_1 = 0. \tag{6.37}$$

   Substituting $\lambda_1 = \lambda^2$, $\lambda_2 = 2\lambda - \lambda^2$ into Equation 6.37, the eigenvalues of system matrix A is given by

$$s_{1,2} = \frac{\lambda_1 + \lambda_2 \pm \sqrt{(\lambda_1 + \lambda_2)^2 - 4\lambda_1}}{2} = \lambda. \tag{6.38}$$

   Since the filter factor $\lambda < 1$, the eigenvalues norm is strictly less than 1. According to the stability of the linear discrete system, if the system input $\alpha(k)$ is bounded, then Equation 6.36 is ISS.

2. Since the two outputs $\hat{\alpha}(k)$, $\hat{\dot{\alpha}}(k)$ are ISS, the filter convergence can be analyzed by two steps. First, from Equation 6.36, we can see that

$$\hat{\alpha}(k) = (1 - G)\hat{\alpha}(k-1) + (1 - G)T_c\hat{\dot{\alpha}}(k-1) + G\alpha(k) . \tag{6.39}$$

   When $\hat{\alpha}(k)$ reaches its steady state, there exists a sufficiently large integer n_0, $\forall k > n_0$, $\hat{\dot{\alpha}}(k) \to 0$, and $\hat{\alpha}(k) = \hat{\alpha}(k-1)$. Substituting these two conditions into Equation 6.39, we can obtain that $\hat{\alpha}(k) \to \alpha(k)$.

Second, from Equation 6.36, the dynamics of $\hat{\dot{\alpha}}(k)$ is given by

$$\hat{\dot{\alpha}}(k) = -\frac{H}{T_c}\hat{\alpha}(k-1) + (1 - H)\hat{\dot{\alpha}}(k-1) + \frac{H}{T_c}\alpha(k) . \tag{6.40}$$

When $\hat{\dot{\alpha}}(k)$ reaches its steady state, there exists a sufficiently large integer n_1, $\forall k > n_1$, $\hat{\alpha}(k-1) \to \alpha(k-1)$, and $\hat{\dot{\alpha}}(k) = \hat{\dot{\alpha}}(k-1)$. Simultaneously, we can see that $(H/T_c)(\alpha(k) - \alpha(k-1)) = H\dot{\alpha}(k-1)$. Substituting these three conditions into Equation 6.40, we can obtain that $\hat{\dot{\alpha}}(k) \to \dot{\alpha}(k)$. ■

6.2.3 Revised Parametric Adaptive Control Law

In Theorem 6.2, the unknown parameter ϑ_3 is assumed to be constant. This assumption is not appropriate due to the variable external loads on two hydraulic actuators. Thus, the aforementioned parametric adaptive backstepping controller should be revised to guarantee the EHS (6.8) convergence.

Theorem 6.3

Assuming that the unknown parameters ϑ_3 is variable but its derivative $\dot{\vartheta}_3$ is bounded, and the other unknown parameters $\vartheta_i (i = 1, 2, 4, 5, 6)$ are constants, if the virtual control variables $\alpha_i \in C^1 (i = 1, 2, 3)$, and their derivatives $\dot{\alpha}_i (i = 1, 2, 3)$ are bounded, then there exists a revised backstepping controller u that guarantees $z_i(t)(i = 1, \ldots, 4) \to 0$, $\tilde{\vartheta}_i(t)(i = 1, \ldots, 6) \to 0$, as $t \to \infty$.

Proof. The parametric adaptive estimation law of $\dot{\hat{\vartheta}}_3$ in Equation 6.18 can be revised as follows:

$$\dot{\hat{\vartheta}}_3 = -k_3 z_2 + \hat{\dot{\vartheta}}_3 + \left|\Delta\dot{\vartheta}_3\right|_{\max} \operatorname{sgn}(\vartheta_3 - \hat{\vartheta}_3), \tag{6.41}$$

where $\left|\Delta\dot{\vartheta}_3\right|_{\max}$ is the bound of the filter error $\Delta\dot{\vartheta}_3$ (i.e., $\Delta\dot{\vartheta}_3 = \dot{\vartheta}_3 - \hat{\dot{\vartheta}}_3$), ϑ_3 is the calculated value from Equation 6.35, and $\hat{\vartheta}_3$, $\hat{\dot{\vartheta}}_3$ are the filter outputs.

Owing to $\dot{\vartheta}_3 \neq 0$, the Lyapunov function V_2 is rewritten as follows:

$$\begin{aligned}
\dot{V}_2' &= -c_1 z_1^2 + z_1 z_2 + z_2 \left(\frac{A_p}{m} z_3 + \frac{A_p}{m}\alpha_2 - \hat{\vartheta}_1 x_1 - \hat{\vartheta}_2 x_2 - \hat{\vartheta}_3 + c_1 x_2 - c_1 \dot{x}_{1d} - \ddot{x}_{1d}\right) \\
&\quad - \frac{\dot{\hat{\vartheta}}_1}{k_1}\tilde{\vartheta}_1 - \frac{\dot{\hat{\vartheta}}_2}{k_2}\tilde{\vartheta}_2 - \frac{(\dot{\hat{\vartheta}}_3 - \dot{\vartheta}_3)}{k_3}\tilde{\vartheta}_3 \\
&= -c_1 z_1^2 + z_2\left(z_1 + \frac{A_p}{m} z_3 + \frac{A_p}{m}\alpha_2 - \hat{\vartheta}_1 x_1 - \hat{\vartheta}_2 x_2 - \hat{\vartheta}_3 + c_1 x_2 - c_1 \dot{x}_{1d} - \ddot{x}_{1d}\right) \\
&\quad - \tilde{\vartheta}_1\left(\frac{\dot{\hat{\vartheta}}_1}{k_1} + x_1 z_2\right) - \tilde{\vartheta}_2\left(\frac{\dot{\hat{\vartheta}}_2}{k_2} + x_2 z_2\right) - \tilde{\vartheta}_3\left(\frac{\dot{\hat{\vartheta}}_3 - \dot{\vartheta}_3}{k_3} + z_2\right).
\end{aligned} \tag{6.42}$$

Substituting Equation 6.41 and $\dot{\hat{\vartheta}}_1$, $\dot{\hat{\vartheta}}_2$ in Equation 6.18 into Equation 6.42, we can see that

$$\dot{V}_2' = -c_1 z_1^2 - c_2 z_2^2 + \frac{A_p}{m} z_2 z_3 + \frac{\tilde{\vartheta}_3}{k_3}(\Delta\dot{\vartheta}_3 - |\Delta\dot{\vartheta}_3|_{\max} \operatorname{sgn}\tilde{\vartheta}_3). \tag{6.43}$$

According to the convergence of the filter, $\Delta\dot{\vartheta}_3$ is bounded by $|\Delta\dot{\vartheta}_3|_{\max}$. This denotes that the last element in Equation 6.43 is less than 0.

Similarly, if the filter errors $\Delta\dot{\alpha}_2$, $\Delta\dot{\alpha}_3$ (i.e., $\Delta\dot{\alpha}_2 = \dot{\alpha}_2 - \hat{\dot{\alpha}}_2$, $\Delta\dot{\alpha}_3 = \dot{\alpha}_3 - \hat{\dot{\alpha}}_3$) are bounded by $|\Delta\dot{\alpha}_2|_{\max}$, $|\Delta\dot{\alpha}_3|_{\max}$, then the virtual control α_3 is revised as follows:

$$\alpha_3 = -\frac{\frac{m}{A_p} z_2 + c_3 z_3 + \hat{\vartheta}_4 x_2 + \hat{\vartheta}_5 x_3 + \hat{\dot{\alpha}}_2 - |\Delta\dot{\alpha}_2|_{\max} \operatorname{sgn} z_3}{\hat{\vartheta}_6 \sqrt{p_s - \tanh(k x_4) x_3}}. \tag{6.44}$$

Substituting Equations 6.44 and 6.24 into $\dot{V}_3$, the Lyapunov function V_3 is rewritten as follows:

$$\begin{aligned}\dot{V}_3' &= -c_1 z_1^2 - c_2 z_2^2 - c_3 z_3^2 + z_3 z_4 \hat{\vartheta}_6 \sqrt{p_s - \tanh(k x_4) x_3} \\ &\quad + \frac{\tilde{\vartheta}_3}{k_3}(\Delta\dot{\vartheta}_3 - |\Delta\dot{\vartheta}_3|_{\max} \operatorname{sgn}\tilde{\vartheta}_3) - z_3(\Delta\dot{\alpha}_2 + |\Delta\dot{\alpha}_2|_{\max} \operatorname{sgn} z_3),\end{aligned} \tag{6.45}$$

where the last element is less than 0.

Finally, if the controller u is redesigned as follows:

$$\begin{aligned}u = &-\frac{T_{sv}}{K_{sv}} c_4 z_4 + \frac{x_4}{K_{sv}} + \frac{T_{sv}}{K_{sv}} \hat{\dot{\alpha}}_3 - \frac{T_{sv}}{K_{sv}} |\Delta\dot{\alpha}_3|_{\max} \operatorname{sgn} z_4 \\ &- \frac{T_{sv}}{K_{sv}} z_3 \hat{\vartheta}_6 \sqrt{p_s - \tanh(k x_4) x_3},\end{aligned} \tag{6.46}$$

then the derivative of V_4 is rewritten as follows:

$$\begin{aligned}\dot{V}_4' = &-c_1 z_1^2 - c_2 z_2^2 - c_3 z_3^2 - c_4 z_4^2 + \frac{\tilde{\vartheta}_3}{k_3}(\Delta\dot{\vartheta}_3 - |\Delta\dot{\vartheta}_3|_{\max} \operatorname{sgn}\tilde{\vartheta}_3) \\ &- z_3(\Delta\dot{\alpha}_2 + |\Delta\dot{\alpha}_2|_{\max} \operatorname{sgn} z_3) - z_4(\Delta\dot{\alpha}_3 + |\Delta\dot{\alpha}_3|_{\max} \operatorname{sgn} z_4) < 0.\end{aligned} \tag{6.47}$$

By the parametric adaptive estimation law (6.41), the sign of the dynamic element $\tilde{\vartheta}_3$ is guaranteed to be negative in $\dot{V}_4'$. Similarly, the signs of z_3 and z_4 become negative by the virtual control (6.44) and the revised backstepping controller (6.47). This iteration controller guarantees $\tilde{\vartheta}_i (i = 1, \ldots, 6)$ and $z_i (i = 1, \ldots, 4)$ converging to zero. ■

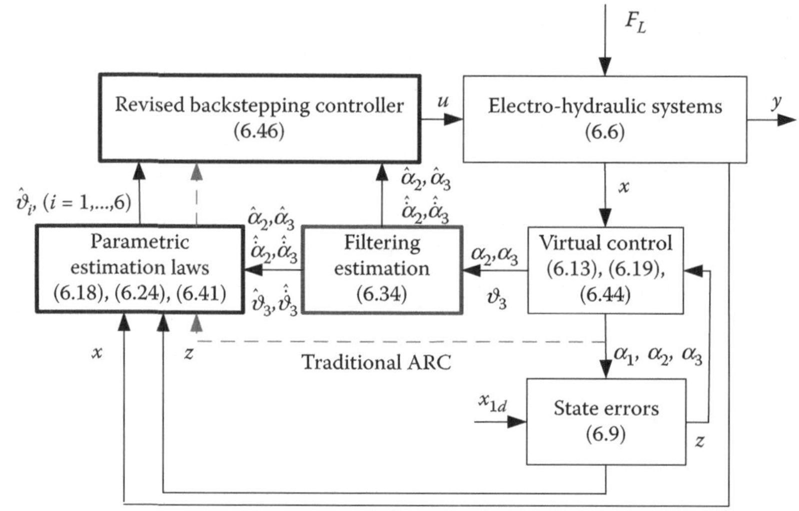

FIGURE 6.1
Block diagram of the control system.

Remark 6.3

The constants $|\Delta\dot{\vartheta}_3|_{\max}$, $|\Delta\dot{\alpha}_2|_{\max}$, and $|\Delta\dot{\alpha}_3|_{\max}$ need to be predefined before the revised backstepping controller (6.46). From Equations 6.35 and 6.40, these three constants can be estimated by $\max_{t\to\infty}\{\dot{\alpha}_i^d(t) - \hat{\dot{\alpha}}_i(t)\}$, $\max_{t\to\infty}\{\dot{\vartheta}_3^d(t) - \hat{\dot{\vartheta}}_3(t)\}$, where $\dot{\alpha}_i^d(t) = |\dot{\alpha}_i|_{\max}\sin 2\pi t$ for $i = 2, 3$, $\dot{\vartheta}_3^d(t) = |\dot{\vartheta}_3|_{\max}\sin 2\pi t$. The constants $|\dot{\alpha}_i|_{\max}$, $|\dot{\vartheta}_3|_{\max}$ can be estimated in the following simplified controller (6.49).

Therefore, the revised backstepping controller (6.46) guarantees $z_i(i = 1,\ldots,4)$ and $\tilde{\vartheta}_i(i = 1,\ldots,6)$ converging to zero. This controller involves the parametric estimation laws (6.18), (6.24), and (6.41), the state errors (6.9), the virtual control laws (6.13), (6.19), and (6.44), and the filtering estimation (6.34) as shown in Figure 6.1.

6.3 Disturbance Observer Application

For another way, the external disturbance $F_L(t)$ can be adopted by the observer disturbance due to the dynamic characteristic of $F_L(t)$. If $f_{21}(x_3) = -A_p x_3/m, f_{22}(x_1) = -x_1/m, f_{23}(x_2) = -x_2/m$, and $d = -F_L(t)/m$ are defined, then the high-gain observer is given by

$$\dot{\hat{d}} = K_d(\dot{x}_2 - f_{21}(x_3) - \hat{\vartheta}_1 f_{22}(x_1) - \hat{\vartheta}_2 f_{23}(x_2) - \hat{d}). \qquad (6.48)$$

In fact, integrating the above equation, the kth disturbance estimation of d is given by

$$\hat{d}(k) = K_d x_2 - \int_{(k-1)T_c}^{kT_c} K_d(f_{21}(x_3) + \hat{\vartheta}_1 f_{22}(x_1) + \hat{\vartheta}_2 f_{23}(x_2) + \hat{d})dt. \tag{6.49}$$

According to Equation 6.8, we can see that

$$\dot{d} = \dot{x}_2 - f_{21}(x_3) - \vartheta_1 f_{22}(x_1) - \vartheta_2 f_{23}(x_2). \tag{6.50}$$

If the disturbance observer error $\tilde{d} = d - \hat{d}$ is defined, then

$$\dot{\tilde{d}} = -K_d\tilde{d} + \dot{d} + K_d\frac{x_1\tilde{\vartheta}_1}{m} + K_d\frac{x_2\tilde{\vartheta}_2}{m}, \tag{6.51}$$

where $\tilde{\vartheta}_1 = \vartheta_1 - \hat{\vartheta}_1$, $\tilde{\vartheta}_2 = \vartheta_2 - \hat{\vartheta}_2$.

Integrating the above equation, we can obtain

$$\begin{aligned}\tilde{d}(t) &= e^{-K_d t}\tilde{d}(0) + \int_0^t e^{-K_d(t-\tau)}\left(\dot{d}(\tau) + K_d\frac{x_1(\tau)\tilde{\vartheta}_1(\tau)}{m} + K_d\frac{x_2(\tau)\tilde{\vartheta}_2(\tau)}{m}\right)d\tau \\ &\leq e^{-K_d t}\tilde{d}(0) + \frac{|\dot{d}|_{\max}(1 - e^{-K_d t})}{K_d} + K_d\int_0^t e^{-K_d(t-\tau)}\frac{x_1(\tau)\tilde{\vartheta}_1(\tau) + x_2(\tau)\tilde{\vartheta}_2(\tau)}{m}d\tau.\end{aligned} \tag{6.52}$$

If the designed parametric adaptive estimation law Equation 6.18 can guarantee two estimation errors ϑ_1, ϑ_2 bounded where the corresponding boundedness is arbitrarily small, then Equation 6.52 can be furthermore handled. In other words, $\forall\epsilon > 0$, $\exists t_1 > 0$, as the time $t > t_1$, $\vartheta_i < \epsilon$ for $i = 1, 2$. We can assume that $\tilde{\vartheta}_i \leq |\tilde{\vartheta}_i|_{\max}$. Thus, Equation 6.52 satisfies the following condition:

$$\begin{aligned}\tilde{d}(t) &\leq e^{-K_d t}\tilde{d}(0) + \frac{|\dot{d}|_{\max}(1 - e^{-K_d t})}{K_d} + K_d\int_0^{t_1} e^{-K_d(t-\tau)}\frac{x_1(\tau)\tilde{\vartheta}_1(\tau) + x_2(\tau)\tilde{\vartheta}_2(\tau)}{m}d\tau \\ &\quad + K_d\int_{t_1}^t e^{-K_d(t-\tau)}\frac{x_1(\tau)\tilde{\vartheta}_1(\tau) + x_2(\tau)\tilde{\vartheta}_2(\tau)}{m}d\tau \\ &\leq e^{-K_d t}\tilde{d}(0) + \frac{|\dot{d}|_{\max}(1 - e^{-K_d t})}{K_d} + \frac{(|x_1|_{\max}|\tilde{\vartheta}_1|_{\max} + |x_2|_{\max}|\tilde{\vartheta}_2|_{\max})e^{-K_d t}}{m} \\ &\quad + \frac{\varepsilon(|x_1|_{\max} + |x_2|_{\max})}{m}.\end{aligned} \tag{6.53}$$

Therefore, as $t \to \infty$,

$$\tilde{d}(t) \leq \frac{|\dot{d}|_{\max}}{K_d} + \frac{\varepsilon(|x_1|_{\max} + |x_2|_{\max})}{m}. \tag{6.54}$$

The large observer gain K_d can reduce the DO error $\tilde{d}(t)$. Furthermore, t_1 is sufficiently large and ϵ is arbitrarily small, which means $\tilde{d}(t)$ is arbitrarily reduced by the observer and estimated gains parameters.

Remark 6.4

The DO $\hat{d}$ (6.48) is not different from the parametric adaptive estimation law ϑ_3, since this DO is directly designed based on the state equations. However, the parametric adaptive estimation law is derived in the backstepping controller iteration.

6.4 Experiment

In this chapter, the two-DOF robotic arm is employed to implement and test the performance of the proposed control method. Meanwhile, several working conditions will be considered when both robotic arms are driven simultaneously, or either one joint is run instead. The specific parameters and brands of the main components in the experimental architecture are listed in Table 6.1.

Some hydraulic and mechanical parameters of this EHS are shown in Table 6.2. Since the hydraulic parameters C_d, w, ρ, β_e, K, b, and C_{tl} are not obtained exactly, which have some uncertainties in different experimental conditions, it is necessary to estimate the uncertain parameters

TABLE 6.1

Specific Parameters and Brand of Main Components

Element	Type	Marks	Quantity
Servo motor	BSM63N-375	BALDOR	1
Fixed displacement pump	TFH-315	Takako	1
Servo valve	D633-R02K01M0NSM2	Moog	2
Hydraulic cylinder	LB6-1610-0080-4M	Hoerbiger	2
Relative encoder	AEDA-3300-BE1	AVAGO	2
Pressure sensor	M3041-000006-350BG	MEAS	4

$\vartheta_i (i = 1, 2, 4, 5, 6)$ by the proposed parametric adaptive estimation law. However, the approximate values $\bar{C}_d, \bar{w}, \bar{\rho}, \bar{\beta}_e, \bar{K}, \bar{b}$, and $\bar{C}_{tl}$ can be preset from some hydraulic references as shown in Table 6.2. If the proposed parametric adaptive estimation law is not adopted, the approximate values $\bar{\vartheta}_i (i = 1, 2, 4, 5, 6)$ without considering parametric uncertainties can also be used in the backstepping controller instead of the estimation values. From the parametric definition in Equation 6.8, the known nominal values are given by

$$\begin{cases} \bar{\vartheta}_1 = \bar{K} \\ \bar{\vartheta}_2 = \bar{b} \\ \bar{\vartheta}_4 = \dfrac{4\bar{\beta}_e A_p}{V_t} \\ \bar{\vartheta}_5 = \dfrac{4\bar{\beta}_e \bar{C}_{tl}}{V_t} \\ \bar{\vartheta}_6 = \dfrac{4\bar{\beta}_e \bar{C}_d \bar{w}}{V_t \sqrt{\bar{\rho}}} \end{cases}. \tag{6.55}$$

Thus, the simplified backstepping controller is redesigned as follows:

$$\begin{cases} u = -\dfrac{T_{sv}}{K_{sv}} c_4 z_4 + \dfrac{x_4}{K_{sv}} + \dfrac{T_{sv}}{K_{sv}} \dot{\alpha}_3 - \dfrac{T_{sv}}{K_{sv}} z_3 \bar{\vartheta}_6 \sqrt{p_s - \tanh(kx_4) x_3} \\ \dot{\alpha}_2 = -\dfrac{m}{A_p}(z_2 + \alpha_1 - \dot{x}_{1d}) - c_2 z_3 - c_2 \alpha_2 \\ \qquad + \dfrac{c_2 m}{A_p}(-\bar{\vartheta}_1 x_1/m - \bar{\vartheta}_2 x_2/m - \vartheta_3/m + c_1 x_2 - c_1 \dot{x}_{1d} - \ddot{x}_{1d}) - \dfrac{c_1 m}{A_p} \bar{\dot{x}}_2 \\ \dot{\alpha}_3 = -\dfrac{z_3 - \dfrac{m}{A_p}(c_2 z_2 + z_1) + c_3(\bar{\dot{x}}_3 - \dot{\alpha}_2) + \bar{\vartheta}_4 \bar{\dot{x}}_2 + \bar{\vartheta}_5 \bar{\dot{x}}_3 + \ddot{\alpha}_2}{\bar{\vartheta}_6 \sqrt{p_s - \tanh(kx_4) x_3}} \\ \bar{\dot{x}}_2 = \dfrac{A_p}{m} x_3 - \bar{\vartheta}_1 x_1/m - \bar{\vartheta}_2 x_2/m - \vartheta_3/m \\ \bar{\dot{x}}_3 = -\bar{\vartheta}_4 x_2 - \bar{\vartheta}_5 x_3 + \bar{\vartheta}_6 \sqrt{p_s - \tanh(kx_4) x_3} x_4 \end{cases}. \tag{6.56}$$

The stroke of the cylinder $S_{\max}$ is 79 mm, but in the experiment, the displacement of the cylinder is not more than 58 mm to avoid the boundary collision of the robotic arm.

The initial values of the estimated parameters are predefined as zero, that is, $\hat{\vartheta}_{i0} = 0, (i = 1, \ldots, 6)$ to verify the convergence effectiveness of the parametric estimation laws. Some control parameters are designed as follows:

TABLE 6.2

Hydraulic Parameters Used in Experiments

Parameter	Value	Parameter	Value
$\bar{C}_d$	0.62	$\bar{w}$	0.024 m
$x_{v\max}$	7.9 mm	$S_{\max}$	79 mm
p_s	40 bar	A_p	2.01 cm^2
V_t	1.74×10^{-5} m^3	$\bar{\beta}_e$	2.2×10^8 Pa
K_{sv}	7.9×10^{-4} m/V	T_{sv}	12 ms
$\bar{K}$	1000 N/m	$\bar{b}$	2200 Ns/m
$\bar{C}_{tl}$	2.5×10^{-11} m^3/(s · Pa)	$\bar{\rho}$	800 kg/m^3
m_1	1.772 kg	m_2	0.739 kg
m_f	1 kg	I_1	0.071 kgm^2
I_2	0.015 kgm^2	I_{2f}	0.022 kgm^2
m_{2f}	1.739 kg	m_{1f}	3.511 kg
P_1P_2	0.35 m	P_1P_{m1}	0.16 m
P_2P_{m2}	0.12 m	ε_{m1}	7.9°

1. The constants: $k_1 = 10$, $k_2 = 0.1$, $k_3 = 10^{-3}$, $k_4 = 10^{-6}$, $k_5 = 10^{-9}$, $k_6 = 10^{-5}$
2. The constants: $c_1 = 10$, $c_2 = 1$, $c_3 = 10^{-5}$, $c_4 = 10^3$
3. The constants: $|\Delta\dot{\vartheta}_3|_{\max} = 5 \times 10^3$, $|\Delta\dot{\alpha}_2|_{\max} = 5 \times 10^7$, $|\Delta\dot{\alpha}_3|_{\max} = 0.01$
4. The filter parameters: $\lambda = 0.7$, $T_c = 10^{-3}$ s

6.4.1 Result of the Proposed Controller

In order to verify the proposed parametric adaptive backstepping controller in Equation 6.46, two sinusoidal demands of the cylinder displacement are selected as $x_{1d}^s = 29\sin(0.6\pi t)$ mm and $x_{1d}^e = 29\sin(\pi t)$ mm. The four state responses, six parametric estimations, and two control voltages of the proposed controllers are shown in Figures 6.2 through 6.4.

Figure 6.2a shows the good displacement tracking responses of two sinusoidal demands by the proposed controllers. The load pressures measured by pressure sensors are not more than the supply pressure 40 bar shown in Figure 6.2c. Owing to the larger dynamic load on the shoulder actuator, the motion frequency of the shoulder should be lower than that of the elbow. During time slice (0, 0.8 s), the control saturation arises since the initial tracking error is significant. After 0.8 s, the spool positions of two servo valves do not exceed the saturation $\pm x_{vmax}$, which are similar to the dynamic characteristic of two control voltages u_s, u_e shown in Figure 6.4. In the experiment, the robotic arm motion will generate resonance effect when the tracking tendency of two joint angles approach each other. Especially in some time such

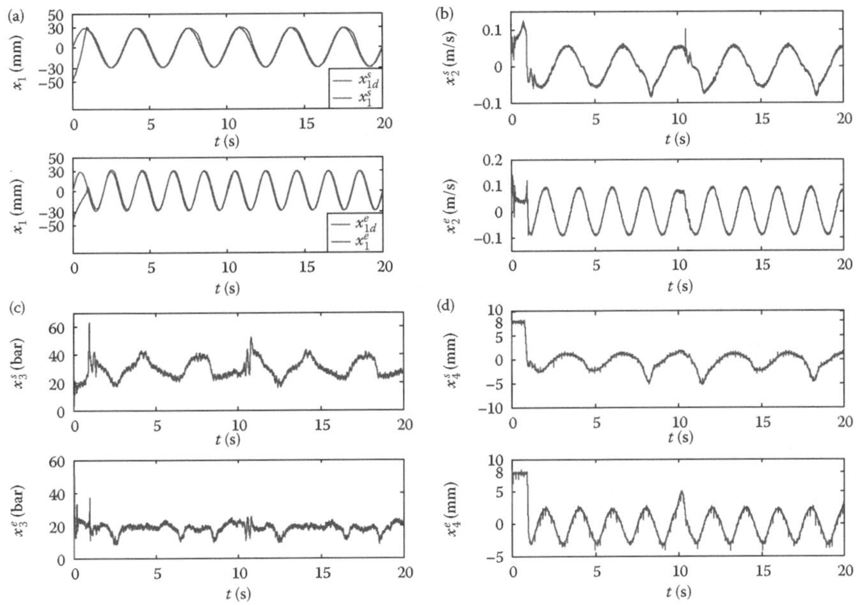

FIGURE 6.2
Four state responses of the EHS by the proposed controller. (a) Cylinder displacement response x_1. (b) Cylinder velocity x_2. (c) Load pressure of cylinder x_3. (d) Spool position of servo valve x_4.

as 8, 11, and 18 s, two hydraulic actuators need simultaneously suffering two maximum external loads; the instant control supplement is significant as shown in Figure 6.4a and b.

The six uncertain parameters are estimated, respectively, by the parametric adaptive estimation laws (6.18), (6.24), and (6.41) shown in Figure 6.3. The parametric estimation $\hat{\vartheta}_1$ approaches the nominal value $\bar{\vartheta}_1$, which denotes that the load spring constant is almost consistent with the approximate preset values. The other four parametric estimations $\hat{\vartheta}_2$, $\hat{\vartheta}_4$, $\hat{\vartheta}_5$, and $\hat{\vartheta}_6$ also converge to the respective steady-state values like $\hat{\vartheta}_1$. However, these parametric estimations have obviously deviated from the approximate preset values, respectively. This phenomenon indicates that the permanent parameters $\bar{\vartheta}_2$, $\bar{\vartheta}_4$, $\bar{\vartheta}_5$, and $\bar{\vartheta}_6$ cannot show the actual hydraulic parameters C_d, w, ρ, β_e, b, and C_{tl} with some uncertainties in different working conditions. Different from the steady characteristic of the other five parameters, $\hat{\vartheta}_3^s$, $\hat{\vartheta}_3^e$ are dynamic estimations, which described two variable external loads of the hydraulic actuators caused by the two-DOF robotic arm motion. If the parametric estimated errors of $\hat{\vartheta}_i (i = 2, 4, 5, 6)$ are significant, the static control bias emerges in long time shown in the initial 1 s in Figure 6.5f. If the dynamic estimations of $\hat{\vartheta}_3$ are inaccurate, the dynamic control saturation emerges in Figure 6.6f.

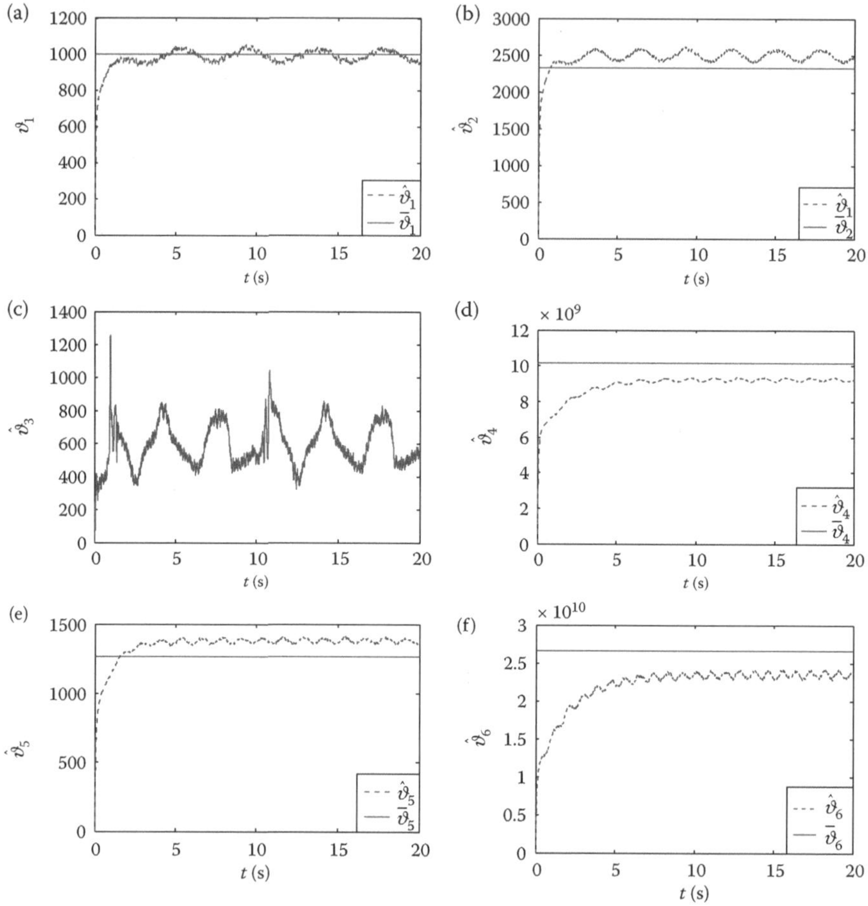

FIGURE 6.3
Six estimation values by parametric adaptive estimation laws. (a) Uncertainty parameter $\hat{\vartheta}_1$. (b) Uncertainty parameter $\hat{\vartheta}_2$. (c) Uncertainty parameter $\hat{\vartheta}_3$. (d) Uncertainty parameter $\hat{\vartheta}_4$. (e) Uncertainty parameter $\hat{\vartheta}_5$. (f) Uncertainty parameter $\hat{\vartheta}_6$.

6.4.2 Comparison with Simplified Backstepping Controller

To illustrate the problem, the comparison results for two different backstepping controllers are given in two critical conditions where the external load of the hydraulic actuator is close to the limitation. The proposed controller (6.46) involves the parametric estimation laws and the filtering estimation. The other simplified backstepping controller (6.56) has no parametric estimation laws and no filtering estimation, where the approximate preset parameters $\bar{\vartheta}_1$, $\bar{\vartheta}_2$, $\bar{\vartheta}_4$, $\bar{\vartheta}_5$, and $\bar{\vartheta}_6$ are determined by Equation 6.55, ϑ_3 is computed by Equation 6.35, without used Equations 6.34 and 6.41. Different from the above experimental condition, the elbow joint is fixed and the

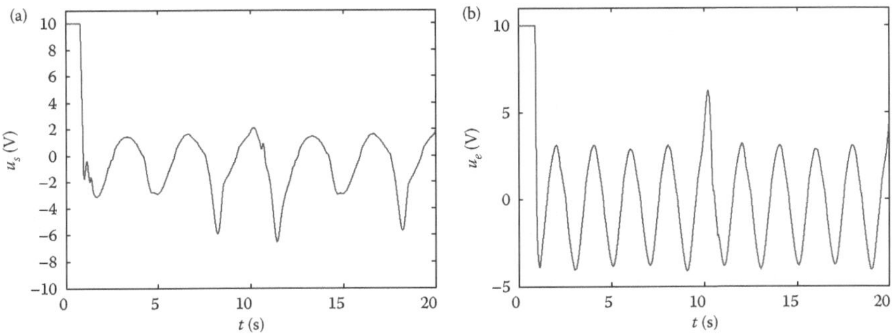

FIGURE 6.4
Control voltages of two hydraulic actuators by the proposed controller. (a) The control voltage of shoulder controller u_S. (b) The control voltage of elbow controller u_e.

displacement demands of shoulder are selected as $x^s_{1d} = 14.5\ \sin(2\pi t)$ mm, $x^s_{1d} = 29\ \sin(\pi t)$ mm. The comparison results are shown in Figures 6.5 and 6.6. The derivatives $\dot{\alpha}_2$, $\dot{\alpha}_3$ obtained by the proposed filter are significantly smaller than the computation values by Equation 6.56 as shown in Figures 6.5a–d. This experimental phenomenon denotes that the proposed filter has more capability to suppress the violent derivative of virtual control than the conservative approach based on model computation. Figure 6.5e shows that the proposed controller has higher tracking performance, since the parametric estimation laws are used to adapt the actual hydraulic parameter with uncertainties. On the contrary, the simplified backstepping controller (6.56) with certain parametric assumptions decline the dynamic behavior of the closed-control loop and results in the sharp controller as shown in Figure 6.5f.

Especially in large stroke motion experiment of the shoulder actuator as shown in Figure 6.6, the simplified backstepping controller could not obtain the satisfactory tracking performance. However, to some extent, the proposed filter still guarantees the dynamic response of the shoulder actuator. These results show that the large stroke motion is easier to cause the saturated and sharp controller than the high-frequency motion due to the dynamic external load on the shoulder actuator.

6.5 Result of Disturbance Observer

The DO (6.48) can replace the estimation law (6.41) since the two external loads on two hydraulic actuators are dynamic variables. In simulation, the external loads F_{L1} and F_{L2} on two EHAs are computed by Lagrange equation using some mechanical parameters of the robotic arm. This two values can

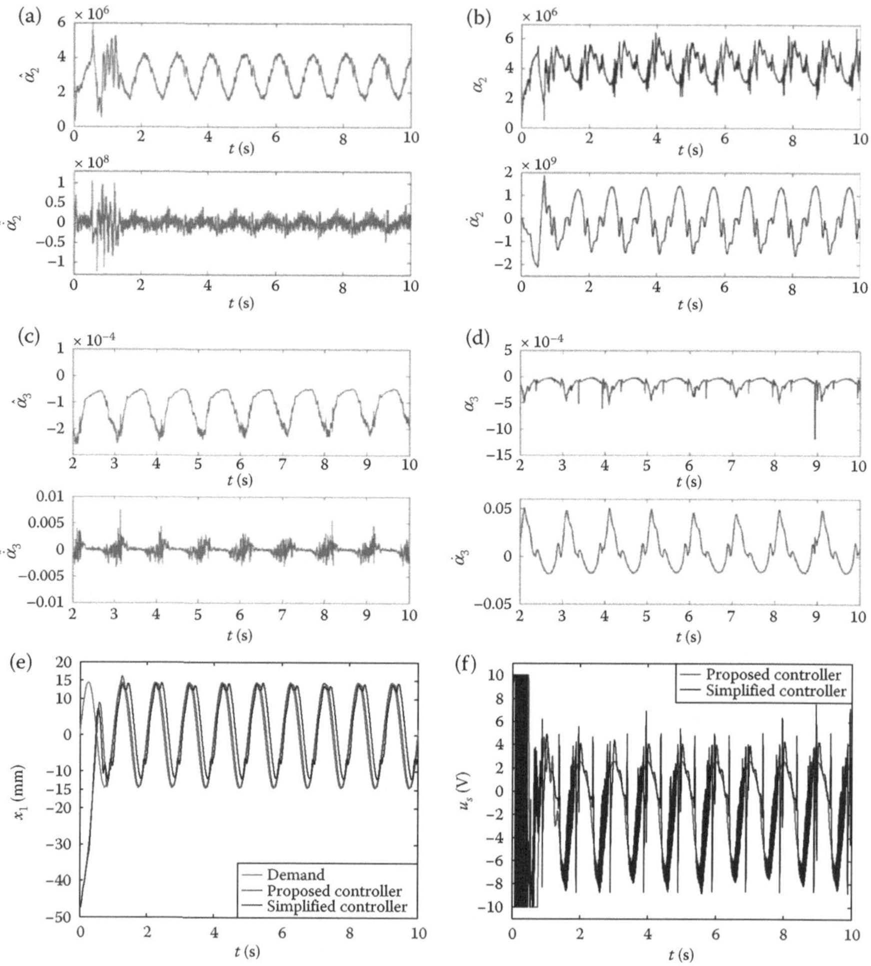

FIGURE 6.5
Comparison result for the demand input $x^s_{1d} = 14.5\sin(2\pi t)$ mm. (a) α_2 and $\dot{\alpha}_2$ by the proposed filter. (b) α_2 and $\dot{\alpha}_2$ by Equation 6.56. (c) α_3 and $\dot{\alpha}_3$ by the proposed filter. (d) α_3 and $\dot{\alpha}_3$ by Equation 6.56. (e) Cylinder displacement response x_1. (f) The control voltage of shoulder controller u_S.

be selected as the demand of the DO. Then the external load disturbances on two EHAs are estimated by the DO (6.48) as shown in Figure 6.7. This high-gain observer guarantees the observer errors $\tilde{d}_1$ and $\tilde{d}_2$ convergence to zero. If the observer error is obvious, the controller will degrade the dynamic tracking performance.

The load disturbance estimations on two hydraulic actuators by the proposed controller are shown in Figure 6.8, which shows that the external

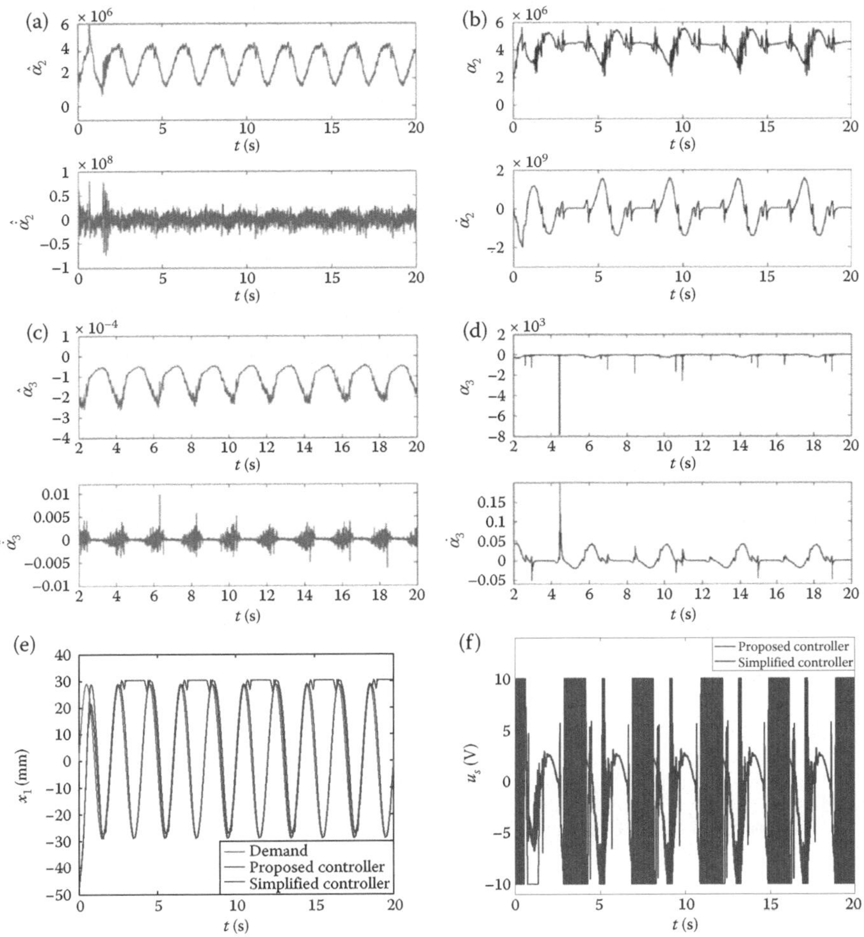

FIGURE 6.6
Comparison result for the demand input $x_{1d}^{s} = 29\sin(\pi t)$ mm. (a) α_2 and $\dot{\alpha}_2$ by the proposed filter. (b) α_2 and $\dot{\alpha}_2$ by Equation 6.56. (c) α_3 and $\dot{\alpha}_3$ by the proposed filter. (d) α_3 and $\dot{\alpha}_3$ by Equation 6.56. (e) Cylinder displacement response x_1. (f) The control voltage of shoulder controller u_S.

load on the upper arm is greater than that on the forearm, although the frequency of the former is smaller than that of the latter. These two disturbance estimations are similar to the simulation results.

The simulation results of position tracking error by two controllers are shown in Figure 6.9. The proposed controller represents the proposed parametric adaptive controller (6.46) with the parametric estimation laws (6.18), (6.24), and (6.41). The other controller combines the simplified backstepping controller (6.56) with the DO (6.48) and the parametric estimation laws (6.18),

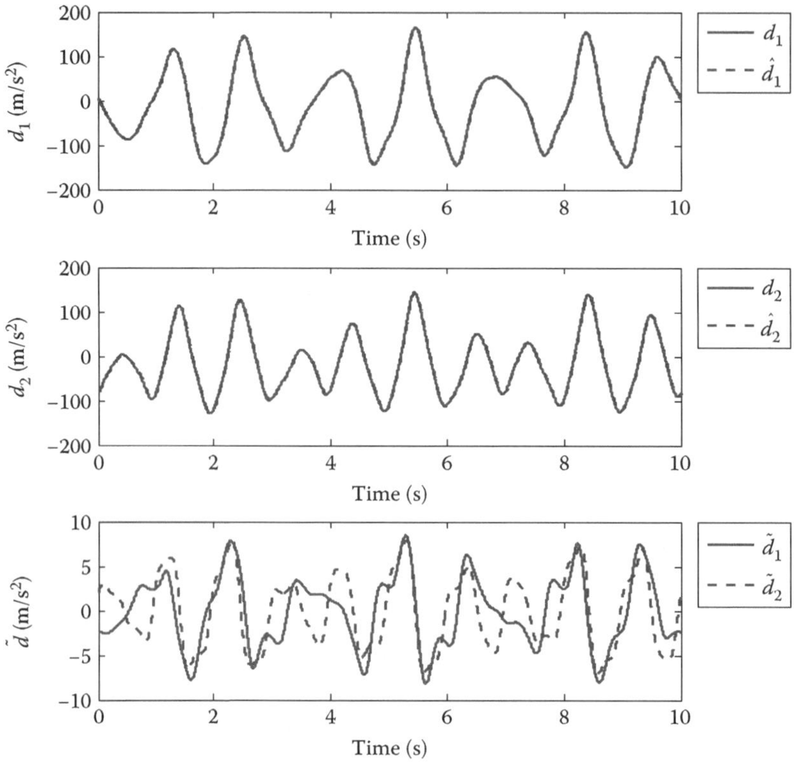

FIGURE 6.7
Simulation results of load disturbance estimation.

(6.24) as follows:

$$\begin{cases} u = -\dfrac{T_{sv}}{K_{sv}}c_4 z_4 + \dfrac{x_4}{K_{sv}} + \dfrac{T_{sv}}{K_{sv}}\dot{\alpha}_3 - \dfrac{T_{sv}}{K_{sv}} z_3 \hat{\vartheta}_6 \sqrt{p_s - \tanh(kx_4)x_3} \\ \dot{\alpha}_2 = -\dfrac{m}{A_p}(z_2 + \alpha_1 - \dot{x}_{1d}) - c_2 z_3 - c_2\alpha_2 \\ \qquad + \dfrac{c_2 m}{A_p}(-\hat{\vartheta}_1 x_1/m - \hat{\vartheta}_2 x_2/m - \hat{d} + c_1 x_2 - c_1\dot{x}_{1d} - \ddot{x}_{1d}) - \dfrac{c_1 m}{A_p}\bar{\dot{x}}_2 \\ \dot{\alpha}_3 = -\dfrac{z_3 - \frac{m}{A_p}(c_2 z_2 + z_1) + c_3(\bar{\dot{x}}_3 - \dot{\alpha}_2) + \hat{\vartheta}_4 \bar{\dot{x}}_2 + \hat{\vartheta}_5 \bar{\dot{x}}_3 + \ddot{\alpha}_2}{\hat{\vartheta}_6\sqrt{p_s - \tanh(kx_4)x_3}} \\ \bar{\dot{x}}_2 = \dfrac{A_p}{m}x_3 - \hat{\vartheta}_1 x_1/m - \hat{\vartheta}_2 x_2/m - \hat{d} \\ \bar{\dot{x}}_3 = -\hat{\vartheta}_4 x_2 - \hat{\vartheta}_5 x_3 + \hat{\vartheta}_6\sqrt{p_s - \tanh(kx_4)x_3}\,x_4 \end{cases} \tag{6.57}$$

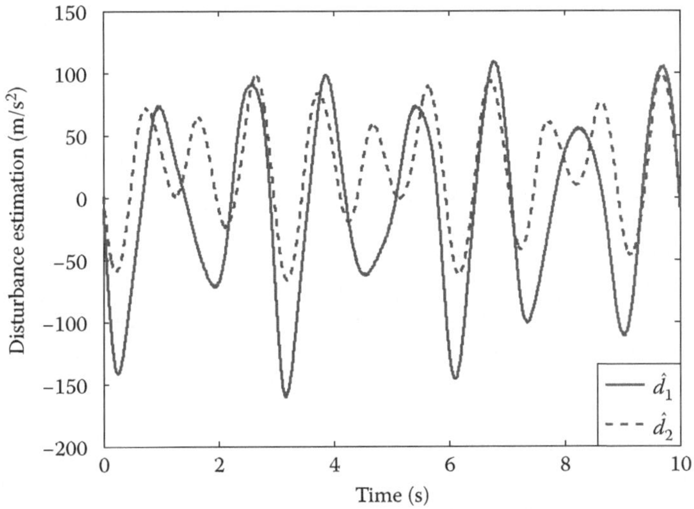

FIGURE 6.8
Experimental results of the load disturbance estimation on two EHAs.

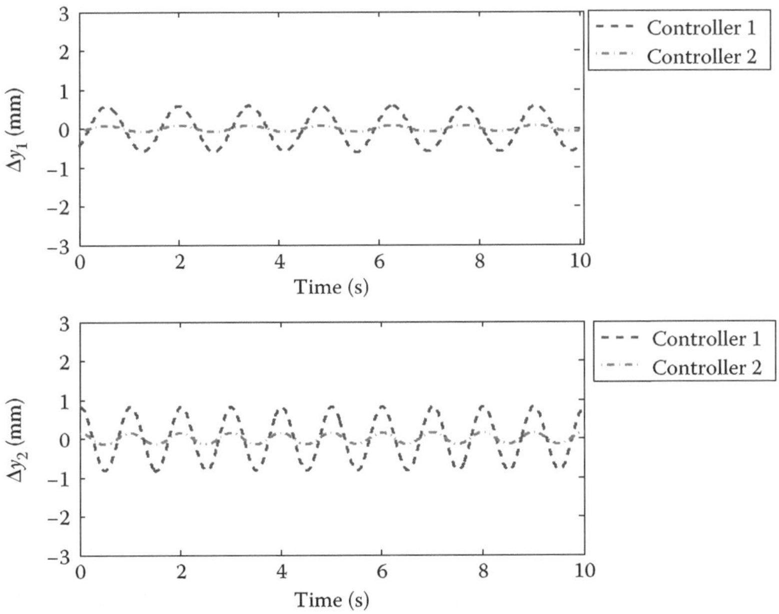

FIGURE 6.9
Simulation results of position tracking error by two controllers, Δy_1—upper arm error, Δy_2—forearm error.

Remark 6.5

It is to be noted that the controller (6.57) is different from the simplified backstepping controller (6.56). In Equation 6.57, the parametric adaptive estimation law can also be used to estimate $\vartheta_i (i = 1, 2, 4, 5, 6)$. Furthermore, the high-gain disturbance observer (6.48) can be adopted to estimate the disturbance $\hat{d}$, which can be directly compensated in the backstepping design.

The performance of the backstepping controller with the DO ($e_{\max} = 0.15$ mm) is better than the proposed controller without the DO ($e_{\max} = 1.13$ mm) as shown in Figure 6.9. This means the DO is constructed to improve the dynamic behavior of the designed controller. The controller can eliminate the bias caused by the unknown dynamic external load.

Figure 6.10 shows that the dynamic tracking performance of the backstepping controller with the DO ($e_{\max} = 1.17$ mm, $\sigma_e = 0.59$ mm) is also better than the proposed controller without the DO ($e_{\max} = 2.89$ mm, $\sigma_e = 1.53$ mm). This result denotes that the tracking performance is declined as the current external load disturbance on the upper arm approaches its limitation ($\hat{d}_1 = 150\,\text{mm/s}^2$) as shown in Figure 6.8. Thus, the backstepping controller with the DO constructs the HGDO to compensate largely dynamic load disturbance and to avoid the obvious error of the position response.

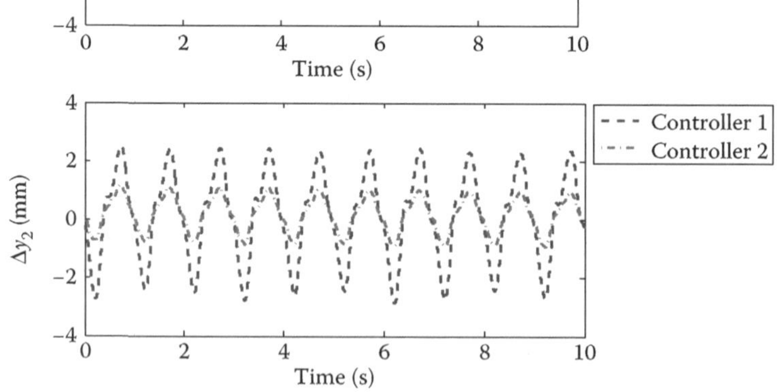

FIGURE 6.10
Experimental results of position tracking error by two controllers, Δy_1—upper arm error, Δy_2—forearm error.

References

1. K. Ahn, D. Nam, and M. Jin. Adaptive backstepping control of an electrohydraulic actuator. *IEEE/ASME Transactions on Mechatronics*, 19(3):987–995, 2014.
2. A. Alleyne and R. Liu. A simplified approach to force control for electro-hydraulic systems. *Control Engineering Practice*, 8(12):1347–1356, 2000.
3. A. Alleyne and R. Liu. Systematic control of a class of nonlinear systems with application to electrohydraulic cylinder pressure control. *IEEE Transactions on Control System Technology*, 8(4):623–634, 2000.
4. G. Balas and J. Doyle. Robustness and performance trade-offs in control design for flexible structures. *IEEE Transactions on Control Systems Technology*, 2(4):352–361, 1994.
5. J. Bobrow and K. Lum. Adaptive, high bandwidth control of a hydraulic actuator. *ASME Journal of Dynamic System, Measurement, Control*, 118(4):714–720, 1996.
6. I. Boiko. Variable-structure PID controller for level process. *Control Engineering Practice*, 21(5):700–707, 2013.
7. F. Bu and B. Yao. Observer based coordinated adaptive robust control of robot manipulators driven by single-rod hydraulic actuators. In *Proceedings of 2000 IEEE-ICRA*, San Francisco, CA, pp. 3034–3039. IEEE, 2000.
8. F. Bu and B. Yao. Nonlinear model based coordinated adaptive robust control of electro-hydraulic robotic arms via overparametrizing method. In *Proceedings of 2001 IEEE-ICRA*, Seoul, Korea, pp. 3459–3464. IEEE, 2001.
9. W. Chen. Disturbance observer based control for nonlinear systems. *IEEE/ASME Transactions on Mechatronics*, 9(4):706–710, 2004.
10. J. Doyle, K. Glover, P. Khargonekar, and B. Francis. State-space solutions to standard H_2 and H_∞ control problems. *IEEE Transactions on Automatic Control*, 34(8):831–847, 1989.
11. C. Du, A. Plummer, and D. Johnston. Variable pressure valve-controlled hydraulic actuation for a robotic arm. In *Proceedings of 9th International Fluid Power Conference*, Aachen, Germany, pp. 186–197, 2014.
12. C. Du, A. Plummer, and D. Johnston. Performance analysis of an energy-efficient variable supply pressure electro-hydraulic motion control system. *Control Engineering Practice*, 48:10–21, 2016.
13. R. Fales and A. Kelkar. Robust control design for a wheel loader using H_∞ and feedback linearization based methods. *ISA Transactions*, 48(3):313–320, 2009.
14. M. Fan, A. Tits, and J. Doyle. Robustness in the presence of mixed parametric uncertainty and unmodeled dynamics. *IEEE Transactions on Automatic Control*, 36(1):25–38, 1991.
15. D. Gu, P. Petkov, and M.M. Konstantinov. *Robust Control Design with MATLAB*. Springer-Verlag, London, 2005.
16. C. Guan and S. Pan. Adaptive sliding mode control of electro-hydraulic system with nonlinear unknown parameters. *Control Engineering Practice*, 16(11):1275–1284, 2008.

17. C. Guan and S. Pan. Nonlinear adaptive robust control of single-rod electro-hydraulic actuator with unknown nonlinear parameters. *IEEE Transactions on Control System Technology*, 16(3):434–445, 2008.
18. Q. Guo, P. Sun, J. Yin, T. Yu, and D. Jiang. Parametric adaptive estimation and backstepping control of electro-hydraulic actuator with decayed memory filter. *ISA Transactions*, 62(S1):202–214, 2016.
19. Q. Guo, T. Yu, and D. Jiang. High-gain observer-based output feedback control of single-rod electro-hydraulic actuator. *IET Control Theory and Applications*, 9(16):2395–2404, 2015.
20. Q. Guo, T. Yu, and D. Jiang. Robust H_∞ positional control of 2-DOF robotic arm driven by electro-hydraulic servo system. *ISA Transactions*, 59(11):55–64, 2015.
21. Q. Guo, Y. Zhang, B. Celler, and S. Su. Backstepping control of electro-hydraulic system based on extended-state-observer with plant dynamics largely unknown. *IEEE Transactions on Industrial Electronics*, 63(11):6909–6920, 2016.
22. Q. Guo, Y. Zhang, and D. Jiang. A control approach for human-mechatronic-hydraulic-coupled exoskeleton in overload-carrying condition. *International Journal of Robotics and Automation*, 31(8):272–280, 2016.
23. W. He, Y. Chen, and Z. Yin. Adaptive neural network control of an uncertain robot with full-state constraints. *IEEE Transactions on Cybernetics*, 46(3):620–629, 2016.
24. W. He, Y. Dong, and C. Sun. Adaptive neural impedance control of a robotic manipulator with input saturation. *IEEE Transactions on Systems, Man, and Cybernetics: Systems*, 46(3):334–344, 2016.
25. W. He and S. Ge. Vibration control of a flexible string with both boundary input and output constraints. *IEEE Transactions on Control Systems Technology*, 23(4):1245–1254, 2014.
26. W. He and S. Ge. Cooperative control of a nonuniform gantry crane with constrained tension. *Automatica*, 66(4):146–154, 2016.
27. W. He and S. Zhang. Control design for nonlinear flexible wings of a robotic aircraft. *IEEE Transactions on Control Systems Technology*, 25(1):351–357, 2017.
28. W. He, S. Zhang, and S. Ge. Adaptive control of a flexible crane system with the boundary output constraint. *IEEE Transactions on Industrial Electronics*, 61(8):4126–4133, 2014.
29. W. He, S. Zhang, and S. Ge. Robust adaptive control of a thruster assisted position mooring system. *Automatica*, 50(7):1843–1851, 2014.
30. C. Johnson and R. Lorenz. Experimental identification of friction and its compensation in precise, position controlled mechanisms. *IEEE Transactions on Industry Applications*, 28(6):1392–1398, 1992.
31. C. Kaddissi, J. Kenne, and M. Saad. Identification and real-time control of an electrohydraulic servo system based on nonlinear backstepping. *IEEE/ASME Transactions on Mechatronics*, 12(1):12–22, 2007.
32. H. Khalil. *Nonlinear Systems* (3rd edition). Prentice-Hall, Englewood Cliffs, NJ, 2001.
33. W. Kim, D. Shin, D. Won, and C. C. Chung. Disturbance-observer-based position tracking controller in the presence of biased sinusoidal disturbance for electrohydraulic actuators. *IEEE Transactions on Control System Technology*, 21(6):2290–2298, 2013.

34. W. Kim, D. Won, and C. Chung. High gain observer-based nonlinear position control for electro-hydraulic servo systems. In *Proceedings of 2010 American Control Conference*, Baltimore, MD, pp. 1440–1446, 2010.
35. W. Kim, D. Won, and C. Chung. Output feedback nonlinear control for electro-hydraulic systems. *Mechatronics*, 22(6):766–777, 2012.
36. W. Kim, D. Won, and M. Tomizuka. Flatness-based nonlinear control for position tracking of electrohydraulic systems. *IEEE/ASME Transactions on Mechatronics*, 20(1):197–206, 2015.
37. M. Krstic, I. Kanellakopoulos, and P. Kokotovic. *Nonlinear and Adaptive Control Design*. John Wiley & Sons, Inc., New York, NY, 1995.
38. G. Liu and S. Daley. Optimal-tuning PID controller design in the frequency domain with application to a rotary hydraulic system. *Control Engineering Practice*, 7(7):821–830, 1999.
39. V. Lu, K. Zhou, and J. Doyle. Stabilization of uncertain linear systems: An LFT approach. *IEEE Transactions on Automatic Control*, 41(1):50–65, 1996.
40. N. Manring. *Hydraulic Control Systems*. John Wiley & Sons, Inc., New York, NY, 2005.
41. H. Merritt. *Hydraulic Control Systems*. John Wiley & Sons, Inc., New York, NY, 1967.
42. V. Milić, Ž. Šitum, and M. Essert. Robust H_∞ position control synthesis of an electro-hydraulic servo system. *ISA Transactions*, 49(4):535–542, 2010.
43. M. Moradi. Self-tuning PID controller to three-axis stabilization of a satellite with unknown parameters. *International Journal of Non-Linear Mechanics*, 49:700–707, 2013.
44. P. Nakkarat and S. Kuntanapreeda. Observer-based backstepping force control of an electrohydraulic actuator. *Control Engineering Practice*, 17(8):895–902, 2009.
45. N. Niksefat and N. Sepehri. Design and experimental evaluation of a robust force controller for an electro-hydraulic actuator via quantitative feedback theory. *Control Engineering Practice*, 8(12):1335–1345, 2000.
46. A. Packard, M. Fan, and J. Doyle. A power method for the structured singular value. In *Proceedings of the 27th Conference of Decision and Control*, Austin, TX, pp. 2132–2137. IEEE, 1988.
47. H. Pan, W. Sun, H. Gao, and X. Jing. Disturbance observer-based adaptive tracking control with actuator saturation and its application. *IEEE Transactions on Automation, Science and Engineering*, 13(2):868–875, 2016.
48. Y. Pi and X. Wang. Observer-based cascade control of a 6-DOF parallel hydraulic manipulator in joint space coordinate. *Mechatronics*, 20(6):645–655, 2010.
49. A. Plummer and N. Vaughan. Robust adaptive control for hydraulic servo systems. *ASME Journal of Dynamic System, Measurement, Control*, 118(2):237–244, 1996.
50. Y. Qiu, X. Liang, and Z. Dai. Backstepping dynamic surface control for an anti-skid braking system. *Control Engineering Practice*, 42:140–152, 2015.
51. C. Semini. The report of production display for "HyQ robot" made in the Dynamic Legged Systems Lab, Istituto Italiano di Tecnologia (IIT), Genova, Italy, http://www.iit.it/en/advrlabs/dynamic-legged-systems.html.
52. X. Song, Y. Wang, and Z. Sun. Robust stabilizer design for linear time-varying internal model based output regulation and its application to an electro hydraulic system. *Automatica*, 50(4):1128–1134, 2014.

53. H. Sun and G. Chiu. Nonlinear observer based force control of electro-hydraulic actuators. In *Proceedings of 1999 American Control Conference*, San Diego, California, pp. 764–768, 1999.
54. D. Swaroop, P. Hedrick, J. Yip, and J. Gerdes. Dynamic surface control for a class of nonlinear systems. *IEEE Transactions on Automatic Control*, 45(10):1893–1899, 2000.
55. I. Ursu, A. Toader, A. Halanay, and S. Balea. New stabilization and tracking control laws for electrohydraulic servomechanisms. *European Journal of Control*, 19(1):65–80, 2013.
56. I. Ursu, F. Ursu, and F. Popescu. Backstepping design for controlling electrohydraulic. *Journal of the Frankin Institute*, 343(1):94–110, 2006.
57. W. Gawronski. Balanced systems and structures: Reduction, assignment, and perturbations. *Control and Dynamic Systems*, 54:372–415, 1992.
58. D. Wang and J. Huang. Neural network-based adaptive dynamic surface control for a class of uncertain nonlinear systems in strict-feedback form. *IEEE Transactions on Neural Networks*, 16(1):195–202, 2005.
59. D. Won, W. Kim, D. Shin, and C. Chung. High-gain disturbance observer-based backstepping control with output tracking error constraint for electro-hydraulic systems. *IEEE Transactions on Control System Technology*, 23(2):787–795, 2015.
60. B. Yao and F. Bu. Adaptive robust motion control of single-rod hydraulic actuators theory and experiments. *IEEE/ASME Transactions on Mechatronics*, 5(1):79–91, 2000.
61. J. Yao, Z. Jiao, and D. Ma. Extended-state-observer-based output feedback nonlinear robust control of hydraulic systems with backstepping. *IEEE Transactions on Industrial Electronics*, 61(11):6285–6293, 2014.
62. J. Yao, Z. Jiao, and D. Ma. High-accuracy tracking control of hydraulic rotary actuators with modeling uncertainties. *IEEE/ASME Transactions on Mechatronics*, 19(2):633–641, 2014.
63. J. Yao, Z. Jiao, Y. Shang, and C. Huang. Adaptive nonlinear optimal compensation control for electro-hydraulic load simulator. *China Journal of Aeronautics*, 23(6):720–733, 2010.
64. H. Yu, Z. Feng, and X. Wang. Nonlinear control for a class of hydraulic servo system. *Journal of Zhejiang University Science*, 5(11):1413–1417, 2004.
65. P. Zarchan. *Tactical and Strategic Missile Guidance*. AIAA, Reston, VA, 1997.
66. J. Zhao, J. Wang, and S. Wang. Fractional order control to the electrohydraulic system in insulator fatigue test device. *Mechatronics*, 23(7):828–839, 2010.
67. K. Zhou, J. Doyle, and K. Glover. *Robust and Optimal Control*. Prentice-Hall Inc., Englewood Cliffs, NJ, 1996.

Index